Silk

Reinventing Nature's Superfibre

SEAN BLAMIRES

Paperback: 978-1-965632-90-1
Hardcover: 978-1-966652-03-8
eBook: 978-1-965632-91-8
Library of Congress Control Number: 2024925869

This book is non-fiction and all work done by actual researchers
has been modified and their contributions cited.

Ordering Information:

Prime Seven Media
518 Landmann St.
Tomah City, WI 54660

Printed in the United States of America

Table of Contents

Acknowledgements

The idea to write the first book, called "Silk: Exploring Nature's Superfibre", originated from a lecture series presented at the Universidad de la República, in Rivera, Uruguay, in 2018. I acknowledge my colleagues Luis Fernando Garcia, Mariangeles Lacava, Marc-Antonio Benamú, Martin Santana, and Carmen Viera, and the attendee students, for making those lectures and the subsequent book possible. I thanked many of my colleagues who inspired and assisted with my silk work in various capacities in my original version of 'Silk'. I do not feel I need to individually thank them all again here. I do nevertheless feel the need to re-iterate the extremely positive influence that a few significant individuals have had on my research and subsequent writing. These include I-Min Tso, for his inspiration early in my exploration into the world of silk research, Michael Kasumovic, for the encouragement and facilities to establish the Spider Silk Research Lab at the University of New South Wales, and my long term collaborators Patricia Flanagan, Aditya Rawal, James Hook, Donald Thomas, Dakota Piorkowski, Patrick Spicer, Xungai Wang, Jian Fang, Rangam Rajkhowa, and Benjamin Allardyce, among others, for sticking with the silk program and believing in the great things we were trying to do. I also thank Madeleine and Amelia Blamires for their endless support.

I am grateful to Niall Doran, Jonas Wolff, Anna-Christin Joel, Brent Opell, Matthew Harrington, Janice Edgerely-Brooks, Alexander Baer, and Jan Hamrsky for permissions to use figures and photographs, and Kirk Huffman (Australian Museum) for sharing stories on traditional spider silk uses.

The idea to update and relaunch my original book came from Josh Williams at Prime Seven Media who has been adamant that the silk story is one worth telling in detail. I am grateful for the faith he has shown.

Why study silk?

*S*ilk is the most fascinating of materials. Evolved over hundreds of millions of years, it is undoubtedly nature's superfibre. Unlike any other known fibre, silk has exceptional strength and elasticity with unique optical, thermal, and other properties. What is more this superfibre is created by many so called "lower" organisms, including spiders, moths, and velvet worms. Each type of silk serves a specific purpose, from constructing a spider's webs to protecting a lacewing's eggs or the larvae of moths, to capturing prey for velvet worms. The delicate yet durable threads of silk reveal a fascinating hierarchy, all crafted from proteins in a way that modern science can only aspire to understand. I will delve into silk's wonders, the organisms that produce it, and the latest breakthroughs in harnessing its potential, exploring how this superfibre is inspiring the development of cutting-edge materials with applications ranging from medicine to engineering. Much of what I will discuss has been the culmination of over 15 years of academic research on silks and silk building organisms. First, however, let me indulge myself and go back to the origins of my research career.

I did not intend at the outset of my journey to work on spider or insect silk at all. Indeed, I initiated my research on silks purely by chance. I entered the PhD program at the University of Sydney in the mid-2000s intent on doing some sort of study on reptile ecology. I went so far as commencing a project on the ecology of river turtles in northern New South Wales, and I was, at the time, content on thinking that I was going to do that for the next three years. Anyway, for a range of reasons that project became logistically difficult to sustain, and I had to shift my thinking toward another project. Having lost almost a year of my studentship pursuing the turtle project, I needed to come up with a new project that could yield quality results quickly and get my program back on track. I, with my PhD supervisors, decided it would be interesting to look at the ecology of the St Andrew's cross spiders (*Argiope keyserlingi*, shown in Figure 1) on the University of Sydney's campus. This made sense as they were in abundance most of the year round in the gardens surrounding the Zoology building on the University's Camperdown campus. Conducting the field observations, accordingly, could be done relatively quickly, easily, and inexpensively. It ticked all the boxes as a project that could yield significant results in no more than three years. There was one problem, however. I knew absolutely nothing about spiders or spider webs at the time. I had not previously read anything about their anatomy, ecology, or anything. I had to first familiarize myself with the relevant concepts and theories as fast as possible to be able to ask some profound research questions for my thesis and then go and devise the appropriate experiments to answer them. Needless to say, I spent several months devouring as much of the literature on the subject as I could access from the University's library. I wrote to and, where and when I could, arranged to meet with many of the authors of the papers I was reading.

FIGURE 1. ST ANDREW'S CROSS SPIDER ON ITS WEB. PHOTO TAKEN BY THE AUTHOR IN SYDNEY, AUSTRALIA.

After consulting with my PhD supervisors and seeking specialist advice, I devised what I thought were some remarkably novel experiments. Essentially, I decided to focus my experimental design not so much on the spiders, but on their webs. After all, according to Dawkin's 'extended phenotype' hypothesis, a spider web should come under the same Darwinian selective pressures as any other phenotypic character[1]. I found this idea profound as it meant I could go into the field, or more accurately take a stroll around the University of Sydney's delightful Camperdown campus, and take a bunch of measurements on spider webs. I then used that information to inform me of all I needed to know about the spider's physiology and foraging, and other behaviours. Using such a seemingly simplistic approach I was able to complete my thesis in a very timely manner and have maintained an intense interest in spider webs and silks since. This interest included running a Graduate level course about silk at the Universidad de la República in Rivera, Uruguay, in 2018. It is the content that I generated to deliver those lectures that is the basis for the book 'Silk: Exploring Nature's Superfibre'.

Something that was apparent upon publishing that book was that much of what was mentioned forged ahead rapidly soon after its publication. Of note is the rapidity of development of new silk-based products, while the field of insect and spider silk genetics has also broadened exponentially. This has prompted me, with some pushing from my publishers, to update it, incorporating some of the very latest research and new concepts. Thus, the renewed title is 'Silk: Reinventing Nature's Superfibre'. The trouble is that by the time this book is out the field will probably advance again. It seems things are really moving in the world of silk. And that is fantastic. Let me, however, start here with some of the basics.

More than a fibre

Something I learned from my initial readings and discussions was that everyone, whether they adore them or are repulsed by them, is fascinated by spiders, spider webs, spider silks, and silk in general. Moreover, almost everyone has a story, anecdote, or legend, to share that they hold closely as truths. Over the years I have heard claims such as "I've seen/heard of spider webs catch a bat (and seen the spider eat it)", and "spider silk is tougher than steel and stronger than Kevlar, and the Military is making bullet proof vests out of it", "If spider silk were 3cm thick it could stop a jumbo jet in full flight", and "scientists have genetically crossed spiders with goats to make spider goats that secrete a kind of silk milk", and so on. I had learned that spider webs and silks, and other animal's silks for that matter, intensely fascinate people. Indeed, I am fascinated by them. I have found them awe-inspiring to research and I am learning something new about them whenever I do another investigation or attend another seminar or conference. Let me tell you just some of what I have learned?

As stated already, the idea for this book sprang from lectures I gave in Uruguay in 2018. In those lectures, as I will do so here, I objectively disseminated the science behind spider and insect silks, webs, and other silken structures, in a digestible way. I hope to inform the lay person about the awe-inspiring depth and breadth of the different silks, just as much as the scientist seeking detailed insights. Some myths and legends around some of the silks and their uses will be confirmed, some refuted, others might be generated anew. More than this, however, I am hoping that I can share my fascination for this amazing material by describing some of its astounding physical and chemical properties, examining its cultural significance, and speculating on some future applications. Ultimately, I aim to inspire present and future researchers, inside and outside the directly associated fields of research, to consider investigating silk as a super material for all kinds of novel practical applications.

What exactly is silk?

Silk is a natural proteinaceous secretion of many kinds of invertebrates. It is usually secreted as a fibre, although it may be secreted as a glue or in other form. Some types of silk fibres are woven into textiles by humans. The fibre is usually secreted into the air as a liquid or gel, but then it rapidly self-assembles into a fibre. Proteins comprise most of the bulk of a silk fibre. Silk fibres are called fibroins or, in the case of spider silk, spidroins (which is derived from spider fibroins)[2]. It has, despite being an essentially transparent material, a shiny, even metallic-like- appearance as a fabric, owing to the fibres being in a flattened triangular, or prism, shape and thus it bends light in a quite peculiar way.

Silks have a density of around 1.3 grams per cubic centimetre (written as $g\,cm^{-3}$). While this is lower than many other natural fibres (e.g. wool or cotton, at around $1.5\,g\,cm^{-3}$), it is greater than many synthetic fibres (e.g. polyester, nylon, at $1.1\text{-}1.2\,g\,cm^{-3}$). Nevertheless, it is much less dense than any of the high performing synthetic fibres, like carbon fibre (at $2\,g\,cm^{-3}$). It is this low density that makes silk products, such as scarves, ties and items of clothing, seem extremely light weight for their size. Despite its low density, silk (particularly that from silkworm moth cocoons and spider draglines) has comparable or even greater strength, elasticity, and toughness, than most natural or synthetic materials.

The significance of these properties I will explain later when I discuss some of the intriguing chemical and physical properties of silk, and when describing uses that exploit these properties. The take home message however is that silk has a particularly low mass per unit volume for a high performing material. It is the hierarchical structure of silk, from its sub-molecular order through to its bulk properties (for which I'll give details later), that enables it to be such a high performing material at such a low density.

While the definitions given above describe silk in a rather scientific way, silk is much more than just a kind of material spun by certain taxa of invertebrates, or a fibre of a particular density and material property, or even a fabric exploited by humans to create yarns, garments, robes, or other such materials from. Silk has shaped human knowledge, philosophy, and history in many profound ways. Silk is synonymous with quality, shininess, and luxuriousness. Advertisers use the term "silky" or "smooth as silk" to demonstrate something of high quality, luscious, or grand in texture (as depicted in Figure 2). Glossy paints are referred to as silk paints because they have that shiny appearance associated with silk products. Athletes may be called 'Silk' or 'Silky' if their skills are of the highest quality and extremely clean*. The most senior lawyers, barristers and counsels are colloquially known as 'silks', because their gowns and robes are traditionally made from silk, as opposed to cotton, the wear of more the junior or lower counsel. Ancient trade routes, the so called "silk roads" (which I examine when discussing silk use) were established between the Far East and Europe, specifically to enable the trade of silks and silk clothing and apparel. Silk cloth was used for painting and scrolls in China as far back as 5000BC. Cities and empires have been formed and collapsed as the control of the silk market has shifted over time. It was the primary instigator for the cross-continent transportation of culture, religion, fashion, ideas, and plagues for over a thousand years. Australian and New Guinean Indigenous people have long used spider silk as fishing lines and nets and woven the threads into bags. The people of Vanuatu created night caps and tunics,

FIGURE 2. THE TERM 'SILK' IS OFTEN GIVEN TO PRODUCTS TO CONJURE IMAGES OF HIGH QUALITY, OR LEAVING BEHIND A SMOOTH OR SHINY FINISH.

as well as the sadistic "smothering hood" within which, it is said, widowed women would be suffocated to death to join their dead husbands in the afterlife. If such things intrigue you, you need to read Eleanor Morgan's book Gossamer Days[3] for detailed depictions of these and other long-standing cultural practices around the world.

There are records of spiders and their silk in the writings of the Greek philosophers for centuries, including the Greek Philosophers Democritus, Aristotle and Aelianus, the latter who wrote about the possibilities in imitating the "work of the spiders". One Greek legend from around 300 BCE is that of Arachne, which describes a woman who challenged the goddess Athena in a weaving contest and was turned into a spider as punishment for her audacity.

Shakespeare referenced spider silk's healing properties in A Mid-Summer Night's Dream. Among modern day silk folklore are the cartoons and movies of Spiderman capturing villains using ultra-tough silken threads and webs, swinging from buildings, and stopping trains in full flight with his ultra-tough silk that he ejects as a filamentous substance from a compartment in his forearm. Added to this are the many ridiculous but entertaining horror movies that spiders and their silks have appeared in recently, with the intention of horrifying viewers with imagery of humans becoming entombed alive for later consumption within silken cocoons spun by giant radioactive or hybrid spiders or insects.

---

* The American basketballer Jamaal Wilkes and Australian footballer Shaun Burgoyne for instance were nicknamed Silk as these athletes were extremely skilful at their sports.

The secrete science of silk

Types and variations evolved

The theory of evolution by natural selection, as described by Charles Darwin, is a remarkable idea as it explains the natural process by which all present and past diversity of life on Earth has come about. The actions of natural selection go further than that, however, as they are responsible for shaping some of the most remarkable natural materials ever made. The super-adhesive byssus threads that tie mussels to a rocky platform are one example. The extremely rubbery resilin in the joints of fleas, which enables them to leap over 20 times their own body height in a single effort, is another. So is the super strength of the concrete-like substance called nacre, found in abalone and oyster shells, which resists the impacts of the harshest waves the ocean can muster. Then there is the plentitude of natural adhesives secreted by plants and animals that can reversibly stick to almost any surface. The extremely tough silks of many invertebrates are also examples of some of natural selection's strongest and most robust material products.

It is well known that Darwin encountered several difficulties and dilemmas when devising his theory. One such dilemma was the phenomenon of very similar features appearing in quite disparate organisms or the reappearance of similar solutions to similar problems. The repeated appearance of wings to facilitate flight among insects, birds, bats, and pterosaurs; the streamlining of bodies for swimming among fish, dolphins, and penguins; and the formation of lens-forming eyes across a range of unrelated animals are some textbook examples of Darwin's dilemma. Darwin thought deeply about this and explained it with a theory that became known as Convergent Evolutionary Theory. The theory suggests that natural selection has a limited number of ways to overcome certain problems, making it inevitable that similar characteristics might emerge in otherwise dissimilar organisms.

The independent development of different kinds of silks to perform similar functions among a wide range of different invertebrates seems to be another fascinating but, in my view, underappreciated example of evolutionary convergence. I thus begin my examination of the wonderment of silks by exploring some of the different silk-producing organisms, detailing the properties of their silks, and how they secrete and use it.

So, which animals do secrete silk? I expect that if you were to ask anybody you know that question, most would respond by mentioning the soft, shiny, luxurious-feeling fabrics that are derived from the threads of domesticated silkworm moths (*Bombyx mori*) or the extremely tough silk threads produced by spiders to build their webs. Nevertheless, there are hundreds of different kinds of natural silks, or something akin to silk, produced by many kinds of invertebrates, including various moths, flies, crickets, bees, wasps, ants, leafhoppers, lacewings, caddisflies, embiopterans (the so-called webspinners), mayflies, silverfish, beetles, stink bugs, thrips, mites, velvet worms, spiders, and molluscs. In many instances, the silks used by each of these animals have a wide range of similar properties, despite many of them evolving (as far as we know) independently. Nevertheless, despite developing remarkably similar silk-secreting glands and secretion physiologies, the properties of some silks substantially differ from others. However, as is often the case with natural phenomena, there are just as many exceptions as there are rules when it comes to the biology of silk secretions.

One thing is for sure, there must be something particularly special about silk for it to evolve independently so many times. A multitude of intriguing propositions have been made to explain how the different silk producing glands arose and whether the same selective mechanisms drove the evolution of the different silks and silk glands. A prominent hypothesis postulated by Harvard Biologist Catherine Craig[5] was that the physiological capacity to produce silk evolved early in a common ancestor of two kinds of invertebrates: the Onychophorans, commonly called the velvet worms, and Arthropods, which includes the insects, crustaceans, millipedes and centipedes, and arachnids (Figure 3 shows the evolutionary tree linking these animals). This idea is rather compelling considering that all organisms found today that produce a true silk* are found among the Onychophorans and Arthropods. Under the hypothesis, silk production could have been lost in the trilobites, most chelicerates (a group that includes the spiders, scorpions and mites), the crustaceans, the myriapods (i.e. centipede and millipedes), and most of the insects, and re-formed independently in a few chelicerates, prominently among them are the spiders and some mites, and many insects. Given this, it is extremely fascinating that so many silk producing animals express and spin similar silk proteins, with, presumably, very similar types of silk producing genes†.

---

*   True silks are those comprising of fibrous proteins, or fibroins.

†   A gene is identified as a specific set of nucleotides, which are abbreviated as A, T, C or G, in a stretch of DNA that code for specific proteins.

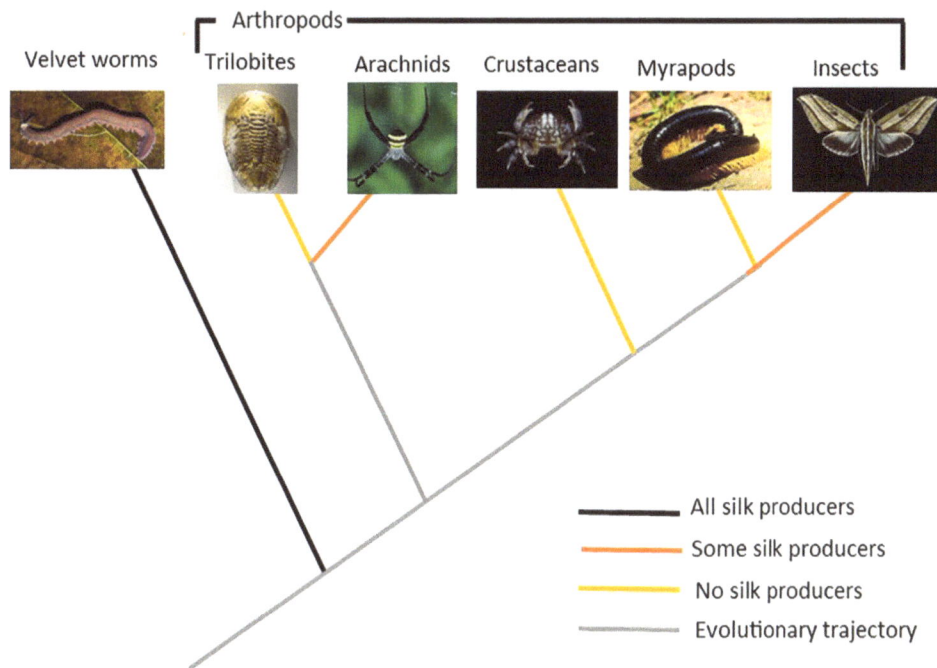

FIGURE 3. INVERTEBRATE EVOLUTIONARY TREE SHOWING THE HYPOTHESIZED EVOLUTIONARY ORIGINS OF SILK PRODUCTION.

A complete overview of the comparative anatomy and behavioral and physiological ecology of some of the most well-researched silk-producing invertebrates is sorely needed. Such a study would no doubt shed light on the factors driving the convergent evolution of silk production among different animals, and help us truly appreciate the selective mechanisms driving silk production and use. This is, nevertheless, not the goal of this book, but it would be fantastic to see such a project undertaken sometime. The goal of this book is to overview what has been researched on the various silk-producing animals and their silks and inspire more research on them to be undertaken. Before proceeding much further, however, I feel compelled to introduce you to some of the vernacular used when describing the structure and functionality of silks, silk proteins, silk fibres, and silk glands at the different hierarchical levels (i.e., from the subatomic to macroscopic). I caution you that I will use some rather technical-sounding terms in this section. Please bear with me as I try to explain these concepts in the simplest of terms possible, as they are crucial for a comprehensive understanding of the subject matter that follows.

A powerhouse of proteins

The silks of most invertebrates, bearing in mind that the exceptions are many, are secreted as fibres that are usually arranged into a hierarchical structure of some kind. These fibres usually consist of a skin or coating comprising lipids (fats), proteins, and/or carbohydrates around a core that primarily consists of long, thin proteins called fibroins or, in the case of spider silk, spidroins. Most fibroins/spidroins are considerably large proteins (>150 kDa, or 150,000 atomic mass units—trust me, this is a darn big molecule) that conform into complex structures. The protein structures themselves are hierarchical in that they form into primary, secondary, and tertiary structures. The primary structure comprises the number, sequence, and bonding arrangements of the amino acid building blocks that make up the protein chains. Since the amino acid sequence is a product of the underlying genes (some basic high school biology is needed here to know about how certain genes are associated with certain amino acids), the amino acid sequences can be used to assess the genetic differences of the silks expressed within and across the different insect or spider species. Most silks contain long stretches of amino acids called domains that may be subdivided into shorter repeating sequences called motifs. Most silk motifs comprise just a few amino acids, most notably the amino acids alanine (usually represented in the scientific literature as 'Ala' or 'A') and glycine (represented as 'Gly' or 'G').

Most of the motifs, and thus most of the domains, are generally highly conserved among the silk-producing invertebrates and form a region in the protein called the repetitive region, or 'repetitive domain.' The repetitive domain may be, in certain silks, subdivided into repetitive motif sets or 'modular units' or 'cassettes.' Other amino acids common in the repetitive domain of silk proteins include serine (Ser or S) and tyrosine (Tyr or Y). It is the terminal ends of the silk proteins, the so-called carboxyl (C-) terminus and the amino (N-) terminus (also called the C- and N-terminal domains), that are non-repetitive (Figure 4). It is in these terminal domains where amino acid compositional variations are more commonly found between species. Now, the primary structure of a silk protein is thought to directly affect the subsequent secondary structure (this, again, requires the reader to understand a bit of protein biology to at least a senior high school level, sorry about that). However, the protein's amino acid sequence is not the only thing that affects its shape. If certain amino acids, such as any ringed and/or kink-shaped amino acids, like tyrosine and proline, are repeatedly incorporated into the protein chains or appear in modified form within the repetitive domain, the protein may twist, turn, and fold in a multitude of ways.

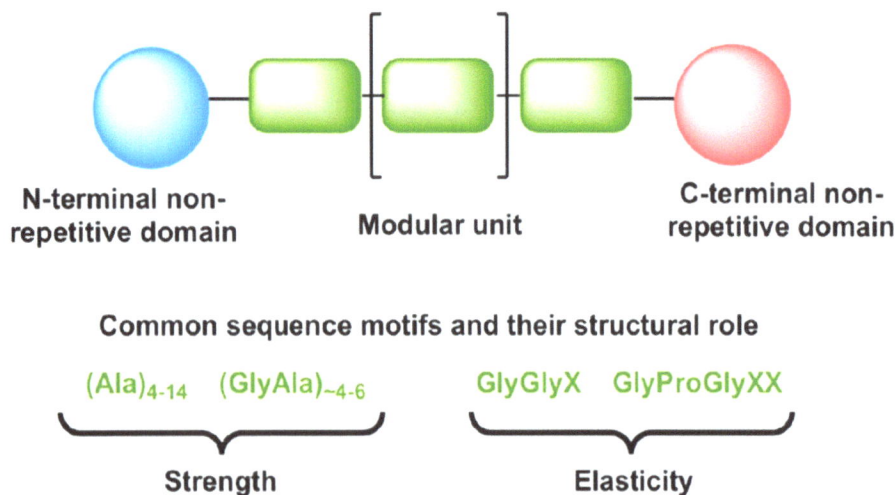

FIGURE 4. THE GENERAL BLUPRINT OF THE MODULAR UNITS AND TERMINAL DOMAINS CHARACTERISTIC OF SILK PROTEINS[7]. THE MODULAR UNITS COMPRISE OF REPEATATIVE AMINO ACID SEQUENCES WHEREAS THE TERMINAL DOMAINS COMPRISE OF NON-REPETATIVE SEQUENCES.

As I have stated, and is accordingly hopefully known, amino acids form the molecular building blocks of proteins. Amino acids are so named because they comprise a base (usually NH2; a nitrogen with two hydrogen atoms), or amino group, at one end of the molecule and a carboxylic acid (COOH; carbon, two oxygens, and a hydrogen), or carboxyl group, attached to the other end, with a so-called 'R' side chain between them. It is this R side chain that is unique in each of the 20 amino acids that form the proteins coded for by the DNA of all living things. Each R side chain has a unique shape, size, charge, and reactivity with other chemicals. The carbon attached to the R side chain is named the α-carbon, with each successive carbon named the β-, γ-carbon, and so on (Figure 5).

FIGURE 5. THE CHEMICAL STRUCTURE OF AN AMINO ACID.

Like all other proteins, the secondary structures of silk proteins (i.e., the various shapes they can become) are formed when certain types of chemical bonds, primarily the weakly interacting bond between the partial charges of different molecular components (these are called hydrogen bonds as they involve hydrogen), form between the amino and carboxyl ends of the constituent amino acids. There are a range of secondary structures that proteins can form, and these have been described as sheets, helices, turns, and coils, among other things. The shapes may be differentiated further according to their length, compression, stacking and folding arrangements, and the strength and positioning of the hydrogen bonds within them. Among large biological proteins, α-helices (formed by the linking up of amino acid-forming α-carbons) and β-sheets appear most commonly. This is also true for silk proteins, with β-coils, β-turns, and shapes called 3_{10}-helices, making up the other common secondary structures formed. The various secondary structures may be stacked or pleated, arranged in a parallel or antiparallel formation, or become cross-linked or randomly arranged. There is a lot of experimental evidence suggesting that the architecture of silk protein secondary structures, and how they interact with each other in the fibre, has very important influences over the mechanical performance of most silks (or how much energy it takes to pull the silk along its axis until it ruptures).

The tertiary structure of a protein is represented by the overall size, shape and conformations of the various secondary structures. In many silks, proteins forming β-sheet secondary structures are stacked into dense units. These units form a so-called crystalline region of the silk. Proteins that have formed β-coils, β-turns, α-helices, 3_{10}-helices, and so on, are arranged into a softer protein matrix known as the amorphous* and lamellar regions or may sometimes just be called the non-crystalline regions. Most silk proteins contain some combination of crystalline, amorphous, and lamellar subunits, with the crystalline units embedded within the amorphous and lamella subunits.

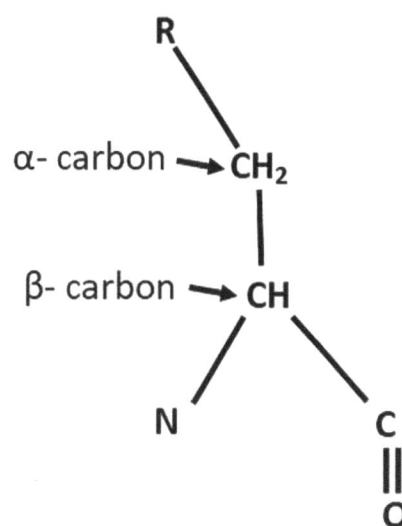

---

* The 'amorphous' region is not truly amorphous as it has structure, so might more accurately be called the 'non-crystalline' region.

The proportion of crystalline subunits compared to amorphous or lamellar subunits is measured as a parameter called the crystallinity index. The crystallinity index is determined using very fancy and sophisticated laboratory measurement techniques such as small and wide-angle X-ray scattering (shortened to SAXS/WAXS, which requires shooting X-rays at the protein molecules and determining how the crystals are arranged from the image the scattered X-rays form on a piece of film), Raman and Fourier Transform Infrared (FTIR) Spectroscopy (which are used to determine protein shape using lasers and infrared radiation, respectively), and is thought to be a crucial indicator of the silk's mechanical performance and thermal properties (that is, how they respond to extreme heat). Some silks might have what could be called semi-crystalline units, but these are less well described in the literature, so I will not mention them again. Researchers have used complex computational modelling techniques such as finite element analysis (FEA) and similar complex computer simulations to identify how the various protein structures interact to influence overall fibre properties, so our picture of how the chemistry of silk influences its performance is now reasonably complete.

As a rule, remembering that there are numerous exceptions, the crystalline β-sheets are tightly packed and so require enormous amounts of energy to rupture their intermolecular hydrogen bonds once they become established soon after the fibre is spun. For this reason, the β-sheets are said to give a particular silk exceptionally high strength. Added to this, the hydrogen bonds that form between amino acids within the silk's amorphous regions also take tremendous amounts of energy to break. However, once they do break, the proteins will slide quite freely past each other whenever the silk is pulled. It is for this reason that most silks have a very high initial stiffness, which is measured by a parameter called Young's modulus, that dissipates away as the fibre continues to be pulled until it ruptures. The point at which the protein-to-protein interactions within the amorphous region weaken to such an extent that the proteins slide freely past each other is called the yield point. How far the proteins slide before the fibre breaks is measured as the extensibility of the fibre and depends on how well aligned the proteins are to each other throughout the amorphous region; the greater the alignment, the less extensible the fibre is.

Silk mechanics: How can it be stronger than steel?

One of the claims I heard very early in my career as an investigator of the properties of spider silks was that "spider silk is stronger than both steel and Kevlar." Is this true? This is a hard one to confirm. I need to explore the intricacies of silk's mechanical properties first. Please bear with me while I do.

Since most silks are secreted as a solid fibre, the properties that are primarily measured as an indicator of silk's material performance are those used to describe fibrous materials, or those measured when the fibre is pulled lengthwise along its axis until it ruptures. The main tensile properties measured by a standard tensile test of a fibre are its: (i) strength, (ii) stiffness, (iii) Young's modulus (also sometimes called the elastic modulus), (iv) extensibility, and (v) toughness. Hardness may be measured by pressing on the fibre to make it rupture. The flexural torsion and a range of thermo-mechanical properties (mechanical properties related to changing temperatures) are sometimes also measured. For silks secreted as liquefied secretions, glues, cement-like materials, or other forms (e.g., foams, films), properties such as viscosity, stickiness, compliance, porosity, compaction, and water solubility are useful to measure. I will describe the properties of fibres just now, leaving discussions of the properties of the various gluey and cementing silks, and the molecular mechanisms by which they are attained, until later.

The way silk fibre tensile properties are generally measured is by fitting or spooling the individual fibres onto cardboard cards and then mounting these cards onto some kind of tensile testing machine, such as an Instron micro-scale tensile tester or a Keysight (formerly Agilent) Universal Testing Machine (UTM), whereupon they are pulled until the fibre ruptures. Many silk fibres are exceptionally thin so require the use of machines with load cells (the sensor allowing the force being applied on the fibre to be measured) with extremely low force sensitivity (much less than 1-Newton, with Newtons being the scientific measure of force). For the determination of properties such as hysteresis, which is also called plastic recovery or the ability to fully return to some initial state, the fibres are not stretched to rupture but to a predetermined load or amount of stretching. Most machines measure the force used to stretch the thread, the load applied, and the extension of the thread. From these measurements a researcher can determine the stress and strain, with stress representing the load on the thread per unit cross sectional area of the thread* and strain as the extension of the thread relative to the thread's initial length. These parameters are then plotted as a stress *vs* strain curve, from which the: (i) ultimate strength (the stress on the thread at rupture), (ii) extensibility (the strain on the thread at rupture), (iii) Young's modulus/initial stiffness, which is taken as the ability of the fibre to initially resist deformation during stretching and is calculated as the slope of the stress-strain curve prior to the yield point, and (iv) toughness (a measure of the energy necessary to rupture the fibre of a given volume), which is calculated as the area under the curve, can be determined[2] (Figure 6).

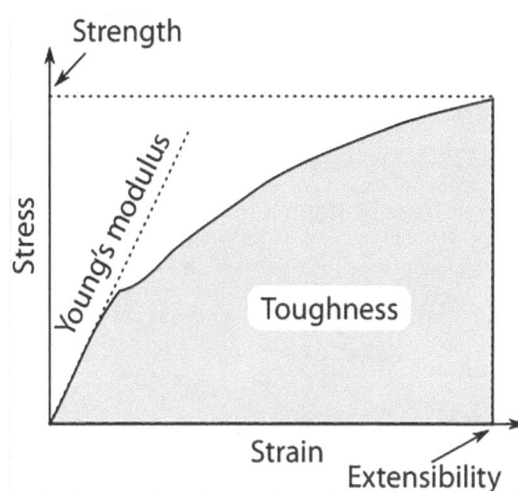

FIGURE 6. STRESS-STRAIN CURVE OF PRODUCED WHEN STRETCHTING A SILK FIBRE, SHOWING HOW THE PARAMETERS STRENGTH, EXTENSIBILITY, YOUNG'S MODULUS, AND TOUGHNESS ARE DERIVED FROM IT.

Two types of measures of stress and strain have been interchangeably used by researchers interested in understanding silk mechanical properties. One is called the 'true' stress and strain and the other the 'engineering' stress and strain[14]. Because silk stretches considerably before it breaks it is likely that it undergoes a change in volume prior to breaking due a phenomenon called fibre crazing, or necking†. For this reason, engineering stress

---

* The cross sectional area of a silk thread is the two dimensional area of the top or bottom of the edge of the thread once it has been sliced. It can be measured by sacrificing a thread for electron microscopy or calculated from thread width by polarizing light microscopy.

† Crazing being the micro-void formation during stretching and necking is the thinning of the fibre when under high strain.

and strain is considered best to use as a measure of silk mechanical performance. Nevertheless, many studies have used true stress and strain and assumed a constant volume across the thread when under different strains[10]. The use of different methods for calculating stress and strains across different studies means that comparisons of the mechanical properties of different silk types among studies either cannot be made with much accuracy or can only be made upon substantial modification of the data. It also means that it was not possible for me to make qualified comparisons of the mechanical properties of the silks produced by different organisms here. I nonetheless have attempted to describe the different uses, protein structures, and some peculiarities, of the silks made by some of the silk spinning organisms.

Not just good looks: silk's optical, thermal and conductive properties

As well as having impressive mechanical properties many silks are very interesting to researchers, and people more broadly, because of their impressive optical properties. I will thus overview these properties here.

One property of silk that is of immense interest is its high refractive index. Refractive index is a dimensionless measure of how much light bends as it travels through a material. It also describes its change in velocity as it moves into the material of interest from the air, or other substance. Air has a refractive index of 1.003, indicating that light passing through it bends or changes velocity only very slightly. Water has a refractive index of 1.33, so light bends considerably more in water than it does in air. While diamond's refractive index is 2.42. The refractive index of silkworm moth silk is about 1.6[11]. That of spider major ampullate (MA) silk is about 1.5[12], which is rather similar to that of glass.

It is this ability to bend a significant amount of the light passing through it that makes silk appear visibly shiny to the naked human eye despite being extremely thin (1-20 μm in width) and otherwise rather transparent. A single thread can also reflect different colours of light when viewed from various angles. These optical properties of silk make it a popular material of choice for the experimental development of organic lenses and sensors, and other such optical devices.

By manipulating the secondary protein structures, researchers can transform silk into non-fibrous materials such as films, hydrogels, and sponges. This adaptability opens up numerous possibilities for the development of biomedical optical, and electronic devices. Lenses, for instance, can be created by purifying the fibroin solution so that it can then be cast into moulds of the desired shape. The concentration and viscosity of the solution are carefully controlled to achieve the necessary thickness and curvature. Silk lenses work by refracting light to focus it in a similar way to conventional lenses made from glass or plastic. Silk lenses may however be better because their refraction may be tuneable. This property could be achieved, as we will see later, through variations in silk protein structure and processing conditions to enable precise control over the silk's properties. Such lenses could be ideal for applications in bioimaging and medical diagnostics as the products could be disposed of and will naturally degrade.

Various types of sensors may also be created from fibroin solutions. However, in this case the silk fibroins get mixed with other functional materials, such as some kind of nanoparticles, enzymes, or fluorescent dyes, to impart specific sensing capabilities. The sensors operate by detecting changes in their environment and converting these changes into measurable signals. For instance, a silken material may be embedded with some kind of fluorescent dye with a sensitivity to changes in charge or pH for the purpose of detecting sections of DNA during genomic analysis or for identifying specific chemicals in the environment.

Naturally spun insect and spider cocoon silks usually appear white, brownish, or beige, but some may appear colourful. Moth cocoon silks, for instance, may appear as brown, green, or pinkish, depending on the moth and/or its diet.* The brown and beige silks from some silkworms are used widely in the textile industry for their versatility at taking up different coloured dyes, thus enabling the creation of a spectrum of fabric colours. The dragline or major ampullate silk and gluey silks of some spiders can appear yellow (Figure 7); some permanently and some intermittently[13]. Other colourful silks include the emerald green silk of spiders of the genus *Ctenophora*, which is prized for some specialty crafts and for scientific research into natural colours. The reddish silk from spiders of the genus *Loxosceles* is known for its antimicrobial properties, and as such has potential applications for generating medical dressings. In certain instances, the colour of an individual spider's or moth's silk may vary according to the time of year they are spun, among other factors[14].

FIGURE 7. THE YELLOW COLOURED SILK OF SOME NATURAL ORB WEBS AND HARVESTED SILKWORM SILK.

---

* A wider range of colours may be attained from genetically engineered moth silks, however.

Silks most often attain their different colours because of the shape and thickness of the skin. The appearance of small cracks in the skin, the accumulation of particulate debris on the surface, or the deposition of pigments within the skin also appear to influence the colour of some spider silks. Since the gain or loss of pigments, such as carotenoids, by animals is often associated with their uptake of the pigments in the diet (as has been identified as the reason for flamingos changing colour as they age), it is likely that variations in dietary uptake is a reason why silk colouration might sometimes vary in individual spiders.

The thermal properties of silk are a set of properties associated with the capacity to absorb temperature. They can be measured in the laboratory by a range of sophisticated techniques with long, often abbreviated names, including thermo-gravitational analyses (TGA), Differential Scanning Calorimetry (DSC), Differential Thermal Analysis (DTA), and Dynamic Mechanical Thermal Analysis (DMTA). The thermal parameters that these techniques allow investigators to determine include enthalpy, glass transition temperature, and melt temperature.

Enthalpy is a measure of the total internal energy dispersal as a function of temperature. The glass transition temperature of a polymer (a long repeating molecule made from combining similar molecules end-on-end) such as silk is the temperature at which it transitions from a stiff/rigid (otherwise called 'glassy') solid to a compliant/ soft ('rubbery') material. In silk the glass transition temperature is usually around 150°C but can vary among different silks and with fluctuations in humidity, particularly among spider dragline silks. The melt temperature is where the rubbery material liquifies and is near 200°C for most silks. The thermal degradation is where the material completely degrades and is usually extremely high. It is an additional parameter that can be measured for silks and other polymers.

TGA is a procedure that determines how the mass of a particular polymer changes because of change in temperature. It enables the researcher to infer properties of the material, including the various phase transitions (such as from a glassy to rubbery phase), enabling the subsequent calculation of glass transition and melt temperatures, energy absorption or desorption, and thermal decomposition temperature. DSC and DTA measure the amount of heat required to increase the temperature of a sample compared to a reference sample, either of air or a substance with known thermal properties. The procedure enables heat flow of the material to be determined, which is directly proportional to enthalpy.[*] DMTA measures the change in mechanical property, usually the loss modulus[†], of a sample as it is heated. Because the thermal parameters measured using each of these procedures correlate strongly with silk's structural features, such as the size, density, and alignment of the crystalline and non-crystalline structures, they are used in complement with small and wide-angle X-ray scattering and similar procedures that estimate the silk protein's secondary structures.

---

[*] According to Kirchhoff's Law the change in enthalpy of a substance is equivalent to its temperature change as long as the pressure of the surroundings is kept constant.

[†] Loss modulus is a measure of a material's capacity to dissipate stresses on it by absorbing heat.

The repetitive nature of the crystalline and non-crystalline secondary structures in most silks enables the transportation of heat and electrical stimuli very effectively. Some silks, such as spider dragline silk, have a conductivity that is greater than some of the well-known conductive polymers, like polyacetylene, and may be comparable to that of carbon nanotubes[15]. Moreover, straining the fibres increases the thermal conductivity, a phenomenon known as a piezoelectric effect, and is the result of a charge building up in the material when under stress. Piezoelectric effects are relatively common among biological materials, especially proteinaceous materials, and are of interest to engineers for producing highly tuned electrical pulses within electronic devices.

Who are the silk makers?

It might surprise you to know that a diverse array of invertebrate animals are silk producers. Among them are velvet worms (animals of the phylum Onychophora), who are close relatives to the Arthropods, which include all insects, the spiders, scorpions, and mites, and crabs and lobsters. These animals excrete sticky silk-like slime from specialized glands to capture prey. Then there are the arthropod silk makers. Among them are the web spinners. These creatures create silk tunnels for protection and habitation, while glow worms use bioluminescent silk to attract prey. Bees, hornets, wasps, and ants utilize silk in nest construction, ensuring structural stability and protection for their colonies. Caddisfly larvae craft underwater silk cases to safeguard themselves against predators and environmental hazards. Many moths, including the domesticated silkworm moth, produce silken cocoons for metamorphosis. It is the latter that humans have harnessed for textile production for millennia. Then there are the spiders, who are well renowned for their diverse silk types, including those used in webs to trap prey, create egg sacs, and create draglines for safety or mobility. Mussels are a mollusc that secretes threads called byssus, which is silk-like but not considered a true silk (as I will explain later) that enables the animal to anchor themselves firmly to surfaces in turbulent environments. As will be explained, each of the following silk-producing organisms has evolved to produce silk for to their specific ecological niches and survival strategies.

Velvet worms: the squishy silk makers

The velvet worms (so named for their velvety appearance and spherical "worm" shaped body) are close living relatives of the Arthropods (the group from which the majority of the remaining silk producing organisms are from), and another group of invertebrates called Tardigrades. Their relationship with Arthropods is such that they are classified within the same major group called Ecdysozoa. This invertebrate group is characterized as having segmented bodies and an exoskeleton that is periodically ecdysed, or shed, as the animal grows. The velvet worms have up to 40 or so paired appendages and can grow up to around 20 cm long.

Velvet worms are considered a living fossil, meaning they closely resemble similar creatures that lived between 419-358 million years ago, a time period known as the Devonian Period. There is anecdotal evidence to suggest that the earliest silk-like secretion was produced by a marine ancestor of velvet worms over 400 million years ago. It is therefore reasonable to predict that the first silks produced might have resembled the offensive secretions used by modern day velvet worms[5]. The kind of proto-silk these animals' use is nonetheless very different from the fibrous silks I had been describing thus far.

Velvet worms are callous hunters, feeding almost exclusively on smaller invertebrate animals. They hunt by squirting out a liquefied silky secretion at any insect it considers will make a nutritious meal. The secretion does not resemble a fibre but is more like a semi-solidified slime (Figure 8). The slime sets hard very quickly and entraps the prey within, holding it in place so that the velvet worm can encase the victim in digestive enzymes so it can then digest it without ingesting it, before sucking up the liquefied remains[16].

FIGURE 8. A VELVET WORM SECRETING ITS SLIMY SILK. IMAGE FROM BAER ET AL. 2017[17] WITH PERMISSION.

Velvet worms squirt their slime out of sets of large, internally branched tubes on its head called oral papillae. These oral papillae are connected to a pair of slime secreting glands that account for over 10% of a velvet worm's body mass. There is a lot of wasted material since it does not always squirt its slime onto its prey with precision. The oral papillae glands are thus massive to ensure that the velvet worm always has a considerable supply of the slimy secretion at the ready. The slime secreting glands are thought to be modified from entirely different glands called crural glands, which are specialized bumps on the legs and body that secrete a range of chemicals in arthropods, including mate attracting pheromones. The crural glands of a particular velvet worm species called *Cephalofovea tomahmontis* are indeed thought to secrete a pheromone as well as the slimy silk-like material[18]. As will become apparent, the way that silk is used among the Arthropods significantly deviates from the way it is used by the Onychophorans, with the Arthropods using it more often for defence, reproduction, shelter, or transport, than attack.

Web spinners: secretive silk secreters

Arthropods of the Order Embioptera are an ancient group of largely tropical insects with specimens very closely resembling modern species being found in amber dating back to the Jurassic Period (i.e. over 145 million years ago). The 360 or so species of modern day embiopterans are found across almost all habitats on Earth and get their common name of "webspinners" from their ability to build expansive silken mats (nicknamed "webs") or galleries under which they permanently retreat from predators, primarily ants, and adverse environments (Figure 9). Their silk-dense galleries can be immense in size and are efficient at protecting the embiopteran builders from attack by ants by producing a physically impenetrable barrier. The arrangement of the silk layers, the way the barrier surface curves with the substrate, and an array of waxy secretions, creates a fully hydrophobic surface across which water and moisture cannot penetrate. While this keeps the webspinners beneath dry during heavy downpours, it means that the webspinners must continuously pierce the gallery to draw in water before rapidly running repairs to these piercings.

FIGURE 9. WEBSPINNER'S GALLERY UNDER CONSTRUCTION. PHOTO BY THE AUTHOR, RIVERA, URUGUAY.

The silks of embiopterans are spun out of thousands of micro-ejectors within their limbs, each of which are associated with an individual silk gland called basitarsomers.[19] The basitarsomers have a relatively simple morphology with two primary structures: a reservoir for storing the pre-spun silk secretions, surrounded by glandular tissue (Figure 10). Prior to spinning the silk secretion is drawn through a filter, the so-called canal cage, into a short spinning duct before being extruded out of the many micro-ejectors.[20]

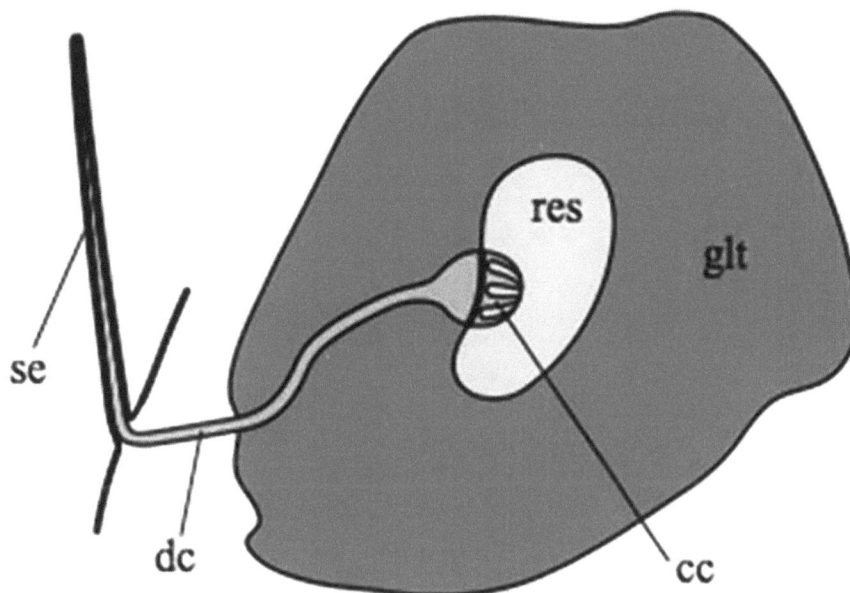

FIGURE 10. ANATOMY OF A WEBSPINNER BARISTOMERE. Se = SILK EJECTOR, dc = THE DUCT, cc = CANAL CAGE res = GLAND RESERVIOR, glt = GLANDULAR TISSUE. REDRAWN FROM BUSSE ET AL. 2015[19] WITH PERMISSION.

Webspinners secrete thousands of incredibly fine silk threads. At no more than 65 nanometres (65 billionths of a metre) across these fibres are considered the thinnest silk threads produced by any animal. Despite the size and number of silk threads and the relatively simple production process, the silk produced by webspinners comprises of highly structured proteins. These proteins are, like many other silk proteins, large and rich in the amino acids glycine and alanine, and analyses have shown that they form thin pleated β-sheet secondary structures.[21] The protein structural arrangement are remarkably like that found in the extremely high performing silks of silkworms and spiders. How such a relatively simple gland and spinning system produces such thin, yet structurally complex, proteinaceous fibres is not understood, and even seems counterintuitive given that the complexity of the gland is generally attributed to the amazing performance of spider silks. It would certainly be worthwhile studying silk production in webspinners in much more detail to get a better understanding of how to go about synthetically producing complex fibres using rather simple apparatus and materials.

Glow worms: shining a light with silk

The name 'glow worm' applies to a range of insects with an ability to produce bright light within their body, a phenomenon called phosphorescence, as larvae or as adults. However, it is the larvae of fungus gnat flies of the largely cave dwelling genus of *Arachnocampa* that produce a very unique type of silk.

There are six known species of *Arachnocampa* glow worm species, all of which are endemic to Australia and New Zealand. The most well-known of which is the New Zealand species *Arachnocampa luminosa*, since it has the most extensive scientific research done on it. This work is primarily on its luminescence, however, and not its silk. These glow worms are found in the popular Waitomo Caves about two and a half hours drive south of Auckland, New Zealand. These caves are visited by over 450,000 people every year. Accordingly, these glow worms are internationally famous. They live the majority of their six to twelve months lifespan as larvae, living in ultra-moist environments such as on the mouth and ceilings of caves and within closed canopy wet rainforests. An individual larva spins a silken retreat from which it will hang, depending on the species, up to 70 silk threads of up to 30 to 40 cm long.

The threads spun by glow worms comprise of a silk fibre encased within a gluey mucous (Figure 11). This mucous seems to rapidly dry out and preventing its drying seems to be a reason why glow worms prefer such ultra-moist environments[29]. The larvae produce a glowing bioluminescence, the production of which is enabled by a glowing pigment called luciferin. The pigment interreacts with the energy carrying molecule Adenosine Triphosphate (ATP) and oxygen, along with an enzyme, luciferase, to produce a startling phosphorescent glow. Being the only source of light within the cave environment, the glow is extremely alluring to a range of insects including mayflies, mosquitoes, caddisflies, midges, moths, and even adult glow worms. These insects fly in a transfixed daze toward the glowing lights before becoming entangled in the sticky silk threads.[22] A mild venom in the mucous is thought to aid in subduing the insect[23]. Soon after an insect becomes ensnared in the sticky trap the glow worm larvae begin to consume the silk threads along with the entangled insect. The phosphorescence is

FIGURE 11. A GLOW WORM THREAD CATCHING A MOTH, FROM PIORKOWSKI ET AL. 2018[22] AND PRINTED WITH PERMISSION.

turned off sporadically while the larvae feed to save its energy. A colony of glow worms within any cave will thus 'sparkle' somewhat like the stars in a starlit, moonless, night sky. It is this effect that makes these animals a major tourist attraction.

After seeing the glowing silken threads of the glow worm *Arachnocampa tasmaniensis* within caves near Hobart, Tasmania, while on an excursion sampling cave spiders, I, along with some collaborators from Macquarie University, Deakin University, Australia, Tunghai University, Taiwan, and the Ludwig Boltzmann Institute in Vienna, Austria, became compelled to initiate a research project on them. Our team first noticed that the silk was very wet and sticky and comprised of an underlying thread smothered in a glue that forms into droplets that superficially look like beads or pearls aligning along a piece of string*.

We also knew of anecdotal reports suggesting that the threads become brittle if dried and the glue loses its stickiness if ambient humidity were to drop even slightly. We therefore thought that glow worm capture silks shared a lot of characteristics with the gluey capture silk that orb web building spiders use to capture their prey (which I will describe in detail later). The first question we asked therefore was whether the capture silks of *Arachnocampa tasmaniensis* and orb web spiders represented an example of convergent evolution and, if so, how far that convergence went? We first tested this idea by performing a series of mechanical performance tests. These initially looked at whether humidity affects the mechanical properties of both the mucous and the axial threads in the same way that humidity affects the glue and underlying axial thread of orb web capture thread silk. A second experiment then compared the chemistry of the gluey mucous with that of orb web capture glues. We followed these up with experiments examining the surface profile and the nano-scale properties of stretched axial threads using some state-of-the-art microscopy techniques with another collaborator from Taiwan.

One functional mechanism that we worked out *Arachnocampa* sticky threads had in common with spider threads was that their adhesion is triggered by water. However, whereas the stickiness of spider sticky threads depreciates gradually as humidity drops, *Arachnocampa* sticky threads are entirely dependent on exposure to a relative humidity of at least 90% to function at all. This, and the physiological requirements of accessible water by the larvae to avoid desiccation, explains their preference for living in or near caves and within wet rainforests.

Another mechanism that *Arachnocampa* threads have in common with spider gluey threads is the way the underlying axial thread elongates when a flying prey animal comes into contact with it. The thread absorbs the impact of the prey striking the thread and prevents the insect from using the thread to springboard back into flight. To accomplish this the underlying axial threads of glow worm capture silks comprises of a pair of very small silk threads working in tandem. In this way the threads differ from spider capture silks, which comprises of a single highly extensible axial thread.

The axial silk itself comprises of proteins that conform into β-sheet secondary structures. These structures are however aligned differently than those in spider capture silk threads. Rather than being aligned parallel with the thread axis, as spider silks are aligned, these are aligned across the axis of the thread in a so-called

---

* Many natural water based materials form this beads on a string arrangement, driven by the internal self-attracting forces of the water molecules, so called Rayleigh Forces. The size and stability of the beads may vary with the temperature and humidity to which they are exposed.

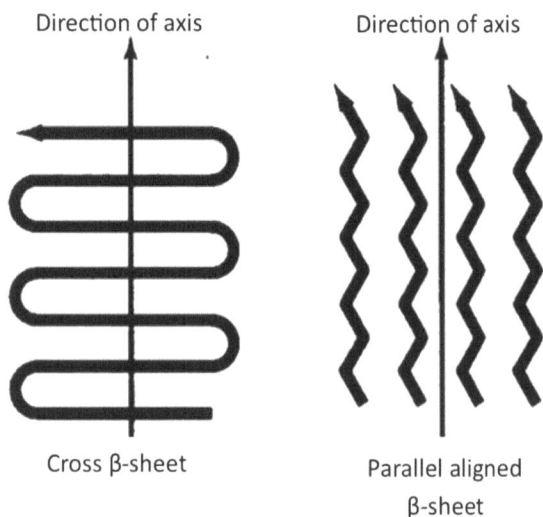

Cross β-sheet   Parallel aligned β-sheet

FIGURE 12. COMPARISON BETWEEN A CROSS-BETA SHEET ARRANGEMENT AND AN ALIGNED BETA-SHEET ARRANGMENT OF PROTEINS IN A FIBRE.

'cross- β-sheet' alignment (Figure 12)[24]. We found nonetheless that like the axial silks of spider capture threads, water affects the mechanical properties of the axial threads. Specifically, the threads become rubbery when wetted. This tells us that, despite the differences in secondary structure, the intermolecular bonding arrangements in the axial silks of both glow worm and spider capture threads must be similar. There are weak hydrogen bonds between the silk protein's amino and carboxyl groups, and these induce the protein chain into slip into alignment along the fibre axis when stretched.

Unlike the glues of spider capture threads, glow worm glues are not anchored to the axial thread in any way. Indeed, when we transported the threads from Hobart to Sydney wrapped around plastic pipette tips, we noticed that the glue would run off the axial thread and accumulate on, or move inside, the small holes at the end of the tips. When we performed chemical analyses on the glues, we found that, apart from a few organic substances such as urea, uric acid, and cresol, there was an entirely different suite of chemicals in glow worm capture glues compared to spider capture glues. The chemicals found included acetate, ethanol, alanine, lactate, methanol, and propylene glycol, among a suite of others. We reran our experiments to compare mainland Australian and New Zealand glow worm glues and found the chemical compositions of the glues from each to differ substantially across species and collection localities[25].

It seems that glow worm glues contain a wider range of compounds of considerably higher molecular mass, such as small proteins and carbohydrates, compared to spider glues[26].* Within glow worm glues these molecules might include venoms for subduing entangled insects and, perhaps, some allomones, which are chemical substances produced and released by a particular species that affects the behaviour of another species, and fluorescent chemicals to lure insects. However, the function of most of the compounds identified remain to be verified. We suspect that these chemicals, along with the fluorescing visual lure emitted from the animal themselves, probably render glow worms very effective and efficient trappers of cave or forest canopy invertebrates.

Our microscopy experiments on the Tasmanian glow worm glues and threads showed there to be significant variation in structure, viscoelasticity, and mechanical performance across individual threads, despite retaining similar surface topologies when the threads were being stretched. We interpreted this as meaning that there may have been a change of structure from cross β-sheets to aligned β-sheets as the fibre became stressed[27], but this hypothesis has not been verified yet by SAXS/WAXS or similar experiments.

---

*   Spider glues contain a wide range of low molecular mass organics molecules. These include, in no particular order: betain, GABA, GABAamide, Choline, Isethionic acid, and N-acetyltaurine. The concentrations of which are highly variable in different spiders, with the humidity of the spider's typical habitat. Glow worm glues do contain these molecules.[p.30]

Raspy crickets: silk spinners with a twist

Raspy crickets are extraordinarily unusual insects in many respects. They belong to the Order Orthoptera, the group comprising the crickets, grasshoppers, locusts, and katydids. Rather than being crickets, they are somewhat an intermediate animal between crickets and katydids. They are named 'raspy crickets' owing to the rasping noises they make as a defence response. Furthermore, they have been reported to pollinate orchids and some other flowers, which is highly unusual for an orthopteran. Even more unusually, raspy crickets, unlike other orthopterans, produce silk. They secrete their silk from an organ called the salivarium into a sac called the acini, whereupon the silk is spun out of the mouth using modified labile glands. No silk micro-ejectors or a set of finger-like spinnerets seem to be present so, essentially, raspy crickets spit out their silk. The silk is used to create shelters under which they hide, and to line their burrows and crevices[35].

Scientists at the Commonwealth Scientific and Industrial Research Organization, CSIRO, in Canberra, Australia, have induced raspy crickets to build large sheets and films of silk in captivity[28]. However, their publication showing this is the only one, to the best of my knowledge, to characterize the structure and function of raspy cricket silk. They found that the silk of raspy crickets is exceptionally fine and comprises of proteins rich in glycine, alanine and serine. These amino acids enable the silk proteins to form pleated β-sheet secondary structures, much like the silks of webspinners, silkworm, and spiders. Accordingly, it is thought to be considerably strong, and the authors even suggest that there is potential to harness the silk for commercial or medical applications using recombinant protein generation. As far as I am aware, however, this has not been accomplished and there has not been any practical applications of raspy cricket silk to date. Raspy cricket silk is nevertheless extremely fascinating, as it represents an example of an independently developed evolutionary convergent silk with a complex molecular structure that is nevertheless secreted by a relatively simple process compared to many of the other types of silk.

Bees and their kin: The triple threat of the insect world - pollinators, predators, silk makers

There are a multitude of hymenopteran insects (that is bees, hornets, wasps and ants) that produce silk for a variety of purposes by a variety of means, yet they remain vastly understudied. Indeed, according to a recent online article, bee silk might just be as commercially useful as silkworm silk.[29] Studies suggest that honeybee silk may be just as strong, or perhaps even stronger, than silkworm silk. Honeybee larvae, like those of glow worms, raspy crickets and silkmoths, produce and secrete silk from modified mouthparts which they use to encapsulate themselves within the hive. Bee silks (and the various wasp and ant silks) achieve great strength in a different way than silkworm or spider silk. Bee silk forms a highly unique protein structure, which is known as a coiled coil arrangement[24]. This unique protein arrangement was independently determined in the 1950s by the renowned biochemists Linus Pauling and Francis Crick, one of the discoverers of the DNA double helix[30,31]. Essentially, it describes proteins that form α-helical structures that intertwine and pack together tightly around each other. The tight packing together of the proteins generates an attractive electrostatic force, called a van der Waals force, between them. This force is difficult to overcome when the thread is being stretched, as such bee silk has considerable strength and initial stiffness.

Hornet (hymenopterans of the genus *Vespa*) queens build hollow nests of masticated tree bark, producing a kind of paper that is divided up into hexagonal cells or combs, into which a single egg is laid. After about a week the egg hatches. Over the next 14 days or so they grow rapidly thanks to the high protein diet they are constantly fed by the queen, whereupon they metamorphose into adult wasps. In many *Vespa* this involves the larvae spinning themselves a case out of silk. The purpose of the silk casing is to provide a constant thermal environment for the metamorphosing pupae. Some studies of the thermal and electrical conducting properties of the silk of the yellow hornet *Vespa orientalis* show it to be quite resistant to electrical charges and heat flow[32]. It thus does an excellent job of protecting the developing hornet pupae. These results also suggest that hornet silk presents as an exceptional material for inspiring the commercial development of light weight thermal insulators and similar materials.

A silk producing gene called Vssilk 5 has been sequenced from the yellow hornet. It was found to code for two cocoon proteins called Vssilk 5 N and Vssilk 5 C. The Vssilk 5 N protein has a non-repetitive amino acid sequence with a central region conforming to a coiled-coil structure. Vssilk 5 C on the other hand consists of serine rich repetitive amino acid sequences and seems to conform into β-sheets. Vssilk 5 C transforms the fibre from primarily α-helix to β-sheet arrangements on stretching. Another silk protein from hornets, Vssilk 5 S, is water soluble and appears to have originated from a diverged segment of Vssilk 5 C[33].

I have observed, along with colleagues at the Universidad de la Republica in Uruguay, several species of wolf spider presenting with the larvae of a parasitic wasp clamped to the side of their abdomen. According to my colleague Luis Fernandez Garcia at the Universidad de la República, in Treinta y Tres, these wasps are of the genus *Paracyphonyx* and *Minagenia* within the wasp family Pompilidae. The usual course of events is for the larvae to continuously suck nutrients and flesh out of the spider over several weeks, eventually rendering the spider little more than an empty spider-shaped exoskeleton. The wasp larvae at this point become so large that it is multiple times larger than the spider. It then detaches from the spider and starts spinning its own web and protective silken barrier around it.

I have not collected quite enough of this silk to yet examine the chemistry or mechanical properties of these wasp's silks, nor has anyone else as far as I know. However, a study of the comparative chemistry of silks of several parasitic wasps from the Family Ichneumonoidea by Donald Quicke and Mark Shaw[34] show there to be remarkable variation in the amino acid compositions across species. Alanine, glycine and serine are consistently found in high compositions, which is consistent with other types of silk, but the occasional finding of high compositions of Asparagine is something very peculiar to these silks. Whether these results mean there is variability in the expression of a silk gene, or genes, across the family is not known, but they suggest it might be the case. Alternatively, the proteins might just have high variability in their composition as a consequence of a varying nutrient composition across the different hosts from which the different species of wasps feed. According to the conventional structure-function models for silks, such variations in amino acid compositions would likely result in high variability in the protein secondary and tertiary structures and silk mechanical properties. Nevertheless, the structural and functional properties are yet to be tested for any wasp silks.

Parasitoid wasps of the genus *Clistopyga* use the silk of its spider host in a rather unusual way. These wasps find jumping spiders inside their silk retreats, then paralyse them using its venom, and insert their long needle-like egg laying appendage, called an ovipositor, through the spider's retreat and into its abdomen to deposit its eggs directly into the spider. The wasp then proceeds to pick at the spider's silken retreat using its ovipositor and resews it with its own silk using in a zigzagging motion of its body to create a more compact case that entraps the spider and its internally developing wasp larvae within[35]. The wasp larvae will grow inside the spider body and devour it from the inside out. All the while it is well protected within its case of silk.

Weaver ants (genus *Oecophylla*) are a group that includes the green tree ants from tropical northern Australia and southern Asia. They are distinctive and large (up to 10 mm long) social ants that build, or 'weave' nests of leaves using silk as the adhesive (Figure 13). The nest is built by the half a million or so workers of the colony. The leaves are bent by strategically chewing the leaf edges and are physically rolled by a team of workers. The leaves are then pasted together with a special glue-like fibrous silk. The silk is, however, is not produced by the workers but by the larvae within the nest. The workers accordingly bring the leaves of the nest together as much as possible and chains of ants form to hold them in their place. The workers then rush into the colony and retrieve the pupating larva. The worker will tap the larva's head to force it to exude silks, using it somewhat like a sealant gun applicator, to patch up the gaps between the leaves.

FIGURE 13. A WEAVER ANT NEST. PHOTOGRAPHED BY THE AUTHOR, PORT DOUGLAS, AUSTRALIA.

Larvae that use up their silk reserves in this process will subsequently pupate without building a cocoon. There have not been any molecular level studies done on harvester ant silks, so we do not know whether it is similar or different to other hymenopteran silks at a structural level or across scales. It does contain a very flexible glue that can withstand the various environmental forces that a nest is exposed to, such as exceptionally strong tropical winds and rainfall. It might thus be reasonable to assume that weaver ant silks have a protein structural arrangement that facilitates high strength, perhaps forming coiled coil structural arrangements in a similar way to that of honeybee silk.

Caddisfly: Riverbed weavers

FIGURE 14. THE CADDISFLY LARVA'S SILK NET. IMAGE MODIFIED FROM http://lifeinfreshwater.net/ WITH PERMISSION.

The insect Order Trichoptera, or the caddisflies, are closely related to the Order Lepidoptera, or the moths and butterflies. Like many of the lepidopterans, the larvae of caddisflies spin a kind of silk. However, since caddisfly live their entire larval period in freshwater, their silks are not used as a retreat during metamorphosis, but for entirely different functions; that being the sticking together of stones and other materials to build shelters and to construct a prey catching net (Figure 14).

The vast differences in the way the evolutionarily related Trichoptera and Lepidoptera use silk gives biologists a special opportunity to compare the roles of evolutionary and ecological requirements in shaping insect silk use and performance, and how protein structures vary with different performance requirements across the different ecological niches of the animals. The ability of caddisfly silk to stick to objects in flowing water, a feat no man-made adhesive comes close to achieving, has prompted many new studies aiming to understand how to produce water-proof adhesives. Accordingly, the way caddisfly use their silk, and its molecular structure and mechanical performance, are becoming reasonably well known.

Caddisfly larvae silk, like lepidopteran silk, is a heterodimer, meaning it is comprised of two or more different types of proteins. These proteins, in the instance of caddisfly silk, are called fibrils. The fibrils are divided into two types, heavy fibrils (or 'H-fibrils') and light fibril (or 'L-fibril'). As with lepidopteran silks (described in detail later), caddisfly silk contains highly repetitive residues of the amino acid cysteine and form into crystalline β-sheet secondary structures[36]. Caddisfly and moth silks are also secreted by morphologically similar spinnerets that are modifications of the respective insect's mouthparts.

There are many significant differences between caddisfly and moth silks, nonetheless. For instance, their amino acid compositions and sequences differ substantially. Unlike moth silks caddisfly silk does not have glycoprotein molecules (molecules comprising of a carbohydrate branch attached to protein base) in the skin. The proline-glycine structures that are highly conserved between and within the silks of caddisfly species are entirely absent from the silk of lepidopterans. But most significantly, caddisfly silk contains a repeating unit abbreviated as $(SX)_4$, with S representing serine and X some other amino acid, most commonly isoleucine or valine. The $(SX)_4$ unit is absent from all lepidopteran silks. Caddisfly silk is, owing to the repeating of $(SX)_4$ units, very high in serine[37]. What is more about 60% of the serine in caddisfly silk is attached to a phosphate ion*. Another way of saying this is to say the serine is "phosphorylated". This creates a substance called phosphoserine. Since phosphorylation of an amino acid usually results in disruption of the protein structure, it remains a mystery to scientists how caddisfly silk remains intact with such a high proportion of phosphoserine.

As the negatively charged phosphate ion attaches to and neutralizes the one end of the serine molecule, the phosphoserine created has a slight negative charge. The positively charged calcium ions present in the silk matrix are exposed along its protein backbone. These enable the protein chains to interact with each other via electrical forces rather than hydrogen bonding, as is the case in other silks. These intermolecular forces induce the protein chains to rigidly align in water. The charged proteins also seem to readily bring about an interaction between the proteins and water, which seems to facilitate reversible adhesion with different surfaces in this type of silk[37].

The rigid alignment of the protein chains are also thought to make the silk exceptionally strong,[38,] enabling them to perform, for some caddisfly species, an additional function of trapping fast moving insect prey and particulate matter as they move within the flowing water at exceptionally high velocities. The architecture of these traps is critically important for their ability to trap particular sizes and types of prey and particles when moving at particular velocities. Accordingly, there is immense variability in trap architectures within and between individual species of caddisfly depending on factors such as the velocity of the water flow and the type of prey and particulate matter present[39].

Lacewings: fiercely protective mothers

Lacewings are a diverse group of insects, with over 1300 species, all belonging to the Order Neuroptera. Species of lacewings vary considerably in size, colour, and behaviour. Unlike most other silk-producing insects, lacewings

---

* Phosphate is a charged particle comprising of multiple elements, also called a radical. It has a positive electrical charge, and so attracts to negatively charged particles, and is represented by the chemical formula $PO4^{3+}$.

produce silk at two stages in their life cycles, as adults when laying eggs, and as final instar larvae to build a cocoon for pupation. Interestingly, the mechanical properties and composition and arrangements of the protein secondary structures differ substantially between the silks used by adult and larval lacewings.

FIGURE 15. IMAGE OF LACEWING EGGSTALK, SHOWING ITS PURPOSE OF PLACING THE EGGSAC SECURELY OUT OF REACH FROM PREDATORY ANTS. FROM EISNER ET AL. 1996[41]; NATIONAL ACADEMY OF SCIENCES, USA.

Lacewings lay their 100 or so eggs individually or in small clusters of two to ten on the underside of plant leaves. To protect these eggs from predators such as aphids and ants the eggs are placed on rigid silken stalks of around 1cm in length. They affix their egg stalks to the leaf using a single drop of a gluey proteinaceous secretion. The silk is secreted as the egg is deposited from the ovipositor as a fluid that rapidly solidifies to form a stable and rigid fibre as a result of exposure to air during post draw elongation. Immediately before the female lacewing deposits her eggs from a tube-like extension from her abdomen, called an ovipositor, that is used solely to deposit eggs, she drops a gluey secretion onto the leaf followed by a droplet of the highly viscous silk. She then moves her abdomen upward which stretches and thins out the viscous droplet to form a thin (less than 15 μm wide) but rigid egg stalk fibre onto which an individual egg sits (Figure 15). Although the thread is exceptionally thin, it has the strength to hold the much more massive egg or egg cluster off the plant surface. Importantly, it has a bending modulus*, three times that of any other silk fibre and the stalk has the capability of immediately recovering to its original position when bent[40].

Lacewings are easy to keep in a laboratory. This, coupled with the incredible flexibility and durability of their silk, make lacewing egg stalk silks ideal for research into high performance fibres and the development of biomimetics. As such, researchers in Bayreuth, Germany[42], have examined the properties of lacewing stalk silks with the intention of harnessing its properties and developing threads for surgical procedures and other medical uses. There is, accordingly, quite a lot of information available about the nanostructures of lacewing egg stalk silk, and how the structures enable such unique mechanical properties.

We know, for instance, that most of the mechanical properties of the egg stalk silk are affected one way or another by the infiltration of water. For instance, the bending modulus increases by a factor of five as relative humidity is increased from 30% to 100%.[24] In this respect lacewing egg stalk silk seems to have some common properties with glow worm silk. Moreover, lacewing egg stalk silk has a cross β-sheet structural alignment that resembles that of glow worm silk[24]. Water softens lacewing silk fibres by disrupting the cross-linking hydrogen bonds between the β-sheets, a trait in common with both glow worm thread axial fibres as well as the axial fibres of spider orb web capture threads. Lacewing egg stalk silk nevertheless shows a unique thinning characteristic called

---

*   Or flexural modulus; it is the capacity to resisting bending when a perpendicular force is applied.

necking when stretched in the direction of the fibre axis. In exhibiting this behaviour, these silks are unlike glow worm threads and resemble some moth or spider silks in which necking occurs to some degree when stretched. It is hypothesized that the cross-β-sheets in lacewing egg stalk silk form into parallel β-sheets as a consequence of elongation during silk deposition[42].

Much less is known about lacewing cocoon silks. It appears, nonetheless, that they share most of their physical and chemical properties with the other insect cocoon silks. For instance, they comprise of exceptionally large fibroins, and the predominant amino acid (over 40% by composition) is alanine, which promotes the formation of pleated β-sheets that align parallel with the fibre axis and render the silk exceptionally strong.

Sea silk: threads from the depths

While not technically a 'silk' as it is not 'spun' from a silk gland nor composed of fibroins or some kind of fibroin-like proteins, sea silk is the name given to the woven cloth fabrics derived from the byssus threads that anchor marine molluscs such as mussels, clams and oysters to a solid substrate. Records of trade in sea silk fabrics and cloth date to ancient Greek and Roman times. The sea silk producers most well studied today are the Mediterranean Pen Shell, *Pinna nobilis*, and the blue mussel, *Mylitus edulis*. These molluscs produce around 60 relatively thick, at about 160 μm or approximately the width of human hair, collagen-like byssal threads. The threads are composed of a combination of keratin (the protein found in our hair and fingernails), and proteins called quinone tanned proteins (these are proteins that help insect cuticle to harden), as well as types of glues, and other proteins. It has impressive material properties quite similar to those of many insect and spider silks.

Bivalve molluscs, which includes the clams and mussels, are so named because they have a hinged, two-part (bivalve) shell made of calcium carbonate. The shells protect their soft body. The main bulk of the body is called the foot and is a muscular organ used for movement or burrowing. The byssus thread is secreted by the foot by contracting to create an air-tight vacuum chamber into which various proteins are pumped as a foam. Continued contractions of the foot manipulates the foam proteins to produce a thread. Finally, a varnish-like substance is secreted across the thread to act as an extremely strong substrate-anchoring glue. Like the true silks, byssus threads self-assemble to form a complex, highly cross-linked, hierarchy of so-called foot proteins (or 'fps')[43].

Byssus foot proteins self-assemble into three vesicle layers consisting of an inner core, a protective cuticle and an adhesive plaque (Figure 16). The proteins of all three layers are linked crossways and contribute to the performance of the thread. It appears that a substance called 3,4- dihydroxyphenylalanine (or, more simply, DOPA) facilitates the cross-linking and plays a role in thread adhesion[44]. The precise method by which the foot proteins are manipulated during fibre formation to form the cross linking nevertheless remains unknown despite there being a considerable interest in these fibres as research subjects for over a century. Indeed, only a few foot proteins like Dpfp2, which stands for *Dreissena polymorpha* foot protein 2 as it is a DOPA containing byssal thread protein isolated from byssus from the zebra mussel *Dreissena polymorpha*, and Dpfp5, an adhesive protein of *D. polymorpha*, have ever been successfully isolated from byssus threads[45]. We know, nevertheless, that a byssus fibre's toughness and its adhesiveness are responsive to changes in the environment they are found in, generally becoming stronger and more adhesive when underwater or as the impacts imparted by tidal forces increase.

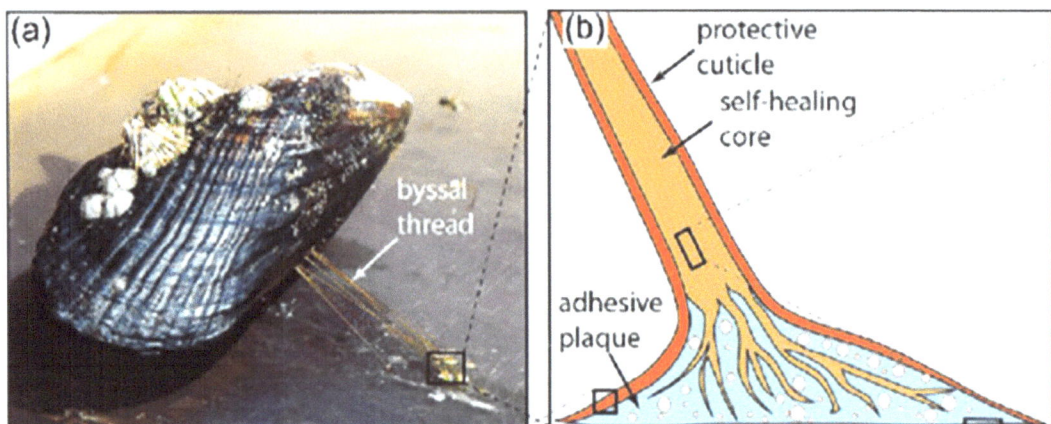

FIGURE 16. PROTECTIVE CUTICLE, CORE, AND ADHESIVE PLAQUE OF MUSSEL BYSSUS THREADS. REDRAWN WITH PERMISSION FROM MATTHEW HARRINGTON.

A byssus thread is subdivided into proximal (near the animal) and distal (distant from the animal) regions. The length of the distal region varies depending on whether or not the animal is fully or partly epifaunal*. The distal region is generally up to four times the length and about half the width of the proximal region. It also has a greater strength and extensibility than the proximal region. One reason for the high extensibility of the distal region is that there is a non-linear relationship between the stress imparted on the fibres and the strain applied. Like moth and spider silk, the typical stress-strain curve of a byssus thread displays a yield point somewhere prior to breaking. At this point the thread becomes re-orientated and shares the applied load along the length of the fibre, effectively increasing the breaking strain experienced along the fibre. Owing to this behaviour, byssus is considered an extremely tough material, albeit with a high degree of variation in its realized toughness depending on the species secreting it, and/or the thread width, and length, and/or the nature of the interactions with the immediate environment. The fact that byssus threads are flexible and retain their functionality on both rough and smooth rocky platforms, and has a consistent high performance in choppy and calm water, and on land, renders it an ideal natural material for inspiring the development of wear-resistant maritime adhesives, and similar applications.

---

* An intertidal animal is said to be epifaunal if it lives permanently on a hard substrate.

Moths: the masters of cocoon crafting

The silk threads from which all of our silk fabrics are made, as well as many medical aids, sutures, tissue cultures, and a suite of other products, are derived by farming, harvesting, and refining cocoon silks of the larvae (a so called a silkworm) of the domesticated Mulberry moth, *Bombyx mori* (belonging to the moth family Bombycidae). *Bombyx mori* is a species of moth that has been selectively bred over millennia to consume mulberry leaves *en mass* as larvae so as to produce vast amounts of silk for human use. It, nevertheless, is not the only moth that produces silken cocoons, nor is it the only one harvested commercially for its silk.

Bombyx mori silk, because of its economic importance and the realization of its unique properties, is by far the most studied silk-producing organism by scientists. It is regarded as the prototype for all silk analyses, including all the silks produced by the insects discussed thus far. It is through studying *Bombyx mori* silk using a wide range of elaborate spectroscopic and other techniques, that researchers know that moth silk has a hierarchical structure with an outer protein coat and inner core of fibroin strands (Figure 17). Four types of protein are found throughout the fibre; a heavy chain fibroin, a light chain fibroin, the so called the H- and L-fibroins respectively, a glycoprotein called P25, and a sticky viscous liquid protein called sericin. The heavy chain fibroin is thought to make up the bulk of the silk fibre and is comprised of highly repetitive GAGAG motifs (and so called the 'gaga' motif) and GAGX motifs, where X can be tyrosine, valine or serine[46]. These amino acids are arranged in a sequence that induces the heavy chain to form a combination of crystalline β-sheets and amorphous β-coils, β-turns, α-helices, and 3_{10}-helices. Studies examining the unique combination of the secondary structures in moth silks have identified that it is the unique combination of crystalline and non-crystalline protein structures in the H-fibroins that give moths' silks their extreme strength and extensibility.

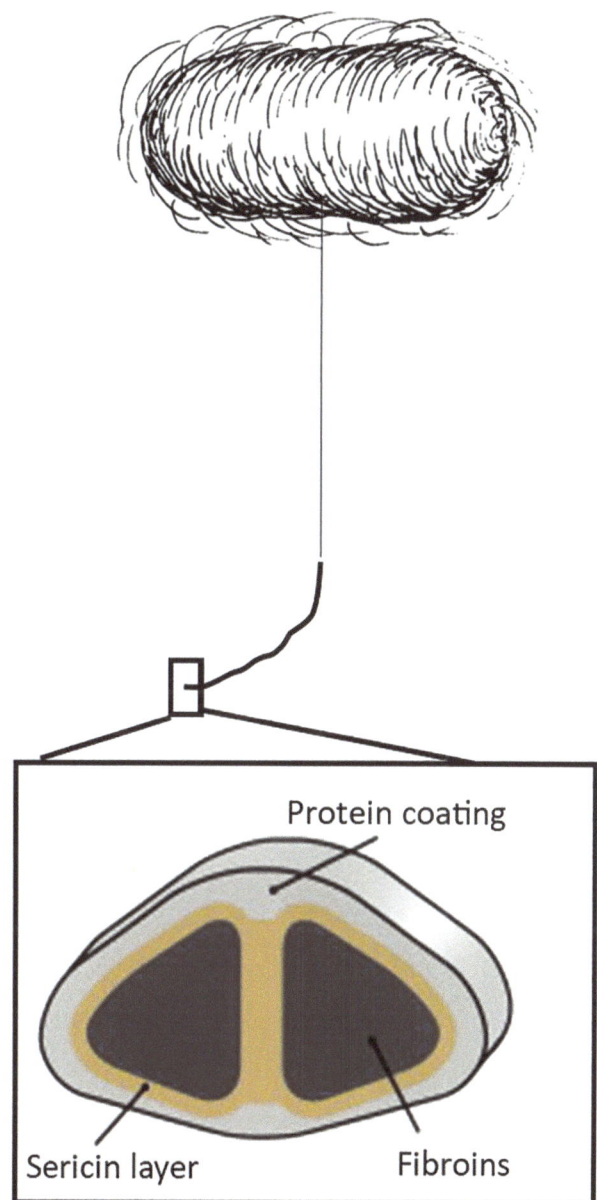

FIGURE 17. STRUCTURE OF SILKWORM SILK, SHOWING THE HIERARCHICAL STRUCTURE AN OUTER PROTEIN COAT, AN INNER CORE AND A SERICIN LAYER.

The P25 protein helps maintain the integrity of the silk core by forming linkages between the H-and L-fibroins. This is accomplished by the hydrophobic glycosylated (a way of saying a carbohydrate has been attached) regions of the P25 protein forming dipole-dipole bonds* and the charged H- and L-fibroins moieties locking the fibroins in place to form a rigid fibroin complex.

Sericin is secreted separately from the bulk fibres by the moth and forms a tacky coating around the threads, which hardens the cocoon and adheres the individual fibres into place. Sericin comprises approximately 20-30% of the cocoon mass. There are three sericin proteins; sericin A, sericin B and sericin C. Unlike other silk proteins, the sericin proteins are composed of a mixture of up to 18 amino acids, with serine the most predominant at around 30%. Most of this serine lies within the C-terminal domain residues of Sericin A.

Separating the serine from the fibrous silk involves boiling the cocoon, a process called "degumming", as the sericin is also called the gum of the cocoon. A detergent might be included to protect the silk threads from damage due to the required high temperature –close to 100°C- of the water. Once removed, the sericin is often discarded, although its proteins can be useful. For instance, their hydrophilic, antioxidant, and ultraviolet reflecting properties render them useful for applications as a hydrating agent within skin cosmetic creams and shampoos, as hydrogels for cell proliferation, and as a natural sunscreen[47].

The use of a cocoon by moths for metamorphosis has repeatedly appeared, disappeared, and reappeared among different moth families over evolutionary time. The distantly related moth families Saturniidae and Bombycidae are the families with the greatest numbers of cocoon-building species among them today. Moreover, the silks from moths of these families are thought to be the toughest, although not all moth silks have been experimentally tested for their strength. Indeed, recent studies on the silk of the bagworm moth (these are larvae of moths belonging to another moth family, the Psychidae) challenge this notion, with claims that their silk may just be as tough, or even tougher, than spider dragline silk[48]. An examination of the bagworm silk genome has revealed unusually large repeats of poly-alanine and glycine-alanine motifs in the repeating regions of its H-fibrils[49]. The genome also reveals that a unique fibroin exists that provides the silk with very high tensile strength. Why this particular type of moth has such exceptionally tough silk is unknown. The cocoon, and indeed the worm inside, can be rather large (up to 15cm long). Sticks and other pieces of debris are embedded throughout the cocoon (Figure 18), so it is extremely heavy. The moth larvae also drag the cocoon around for part of its life. The silks might therefore need to be exceptionally tough in order to protect and suspend such a massive cocoon and larvae off the ground.

---

*   Dipole-dipole bonds are a particular kind of attractive force between charged molecules.

FIGURE 18. PHOTO OF BAGWORM COCOON IN SUBURBAN SYDNEY, AUSTRALIA.

The primary reason speculated for the persistent reappearance of cocoons among different groups of moths is that they serve as protection under extreme climates. This idea seems to go hand-in-hand with observations that tropical moths rarely metamorphose over winter so do not build cocoons. Rather, these moths usually construct a silk-free chrysalis underground. Moths that overwinter in colder climates nonetheless require the protection of a cocoon to maintain their internal temperature for metamorphosis to occur. Again, however, there are many exceptions to these generalizations, and some strange examples of cocoons that appear not to protect the moth larvae from anything at all exist, including the largely transparent cocoon of cabbage looper moths, and the strange meshwork cocoon of the Amazonian urodid moths.

FIGURE 19. THE UNUSUAL MESH COCOON OF THE AMAZONIAN URODID MOTH.

Ignoring the exceptions, the abovementioned generalization seems to suggest it was the thermal properties of cocoons that had been selected for among moths rather than anything to do with the silk's mechanical properties. Indeed, a study my student Hamish Craig did with other researchers at the University of New South Wales, and Deakin University, examined the properties of the silks of three different silkworm moths. These moths were (i) the Mulberry silkworm, (ii) a silkworm called Muga (or *Antheraea assamensis*), and (iii) another called Eri (*Samia cynthia*). The latter two silkworms are members of the moth family Saturniidae. We found that the silks of the latter moths have more and larger nano-sized internal cavities within them than do the silks of Mulberry silkworms. We moreover discovered that the cavities affected the mechanical properties of each of the different types of silks. While most studies attribute the size, shape and conformations of the various secondary and tertiary structures in silkworm silk to most profoundly influence strength, extensibility and toughness of silkworm silk, our work suggests that the size and number of the internal cavities also plays a role. Moreover, we modelled the thermal properties of each of these silks using computer simulations and found that the nano-sized cavities are critical for insulating the cocoons from a harsh thermal environment[50].

We speculated that the larger cavities in Eri and Muga silk existed for temperature regulation inside the cocoon, since these moths live in parts of India that experience great climatic fluctuations, while Mulberry silkworm moths live entirely within artificial indoor chambers at relatively constant temperatures. It also appears that the cavities serve to enhance the performance of the silks, but there seems no apparent ecological reason for this to be so. We expect that cavities within silkworm cocoon silks could supply physical protection and support for the metamorphosing pupae, but studies are needed to test what physical stressors may be encountered by the cocoons and how well the silks can perform under natural conditions.

A more profound dilemma, however, lies in understanding the evolutionary significance of moth silk's extreme toughness. As I have already stated, raspy crickets, bees, caddisfly and lacewing silks have adequately tough silk. It appears that moth silk, owing to its unique combination of protein structures, including a mix of crystalline β-sheets and an amorphous matrix consisting of β-coils, β-turns, and other secondary structures, is however even tougher.

What is it about moth cocoons and the environment they are spun in that necessitates such tough silk? I have alluded to the possibility that natural selection may have directly enhanced the insulating properties of moth cocoon silks rather than mechanical properties. While this might explain why certain moth silks might vary in the size and number of various types of cavities it contains, it does not explain the unique structural properties that have been found in moth silk proteins. I speculate that the silks in moth cocoons, by necessity, need to perform a wider range of functions than do other silks. They need to physically and thermally protect a metamorphosing larva. They also need to support quite a large larva, and the developing adult moth, for substantial periods of time. These functions might only be achievable by using an exceptionally stiff, strong, and tough, silk material.

Spiders: the superstars of silk production

While silk of the domesticated silkworm moth *Bombyx mori* is the most widely recognized thanks to its broad use as a wearable fabric, it is the silken threads spun by spiders that have, as far as we know, the most impressive mechanical properties. Of the 50,000 plus species of extant spider it is the web building species, broadly known as the "true spiders" and scientifically as Araneomorphs, that have the most impressive silk toolkits.

The spider superfamily Araneoidea (often called the "orb web spiders"*) is a large and diverse subfamily of araneomorph spiders that comprises around a quarter of all spider species. A distinctive feature of the Araneoidea is the production of up to seven distinct types of silks, each of which are secreted by a different silk gland (Figure 20). They are the: (i) Major Ampullate (MA) silk, the silk a spider uses to safely fall from its web. It also comprises the radiating spokes and the supportive framework of the familiar wagon wheel-like spider orb webs. (ii) Minor Ampullate (MiA) silk, which is used as a temporary capture spiral thread (the thread going around and around in the orb web to create a barrier into which insects fly) by orb web builders as well as for wrapping prey by a particular phylogenetic group (or clade) of web building spiders called cobweb building spiders[+] (iii) Flagelliform silk, which forms the underlying threads of the capture spirals in orb webs. In orb webs the flagelliform silk is coated with a viscous, sticky, glue. (iv) Aggregate silk is that sticky substance on the capture spirals of orb webs. It is what enables spider webs to retain the insect prey the web intercepts. (v) Pyriform silk is a cement-like silk that holds the web frame to a substrate and adheres different silk threads together during web construction. (vi) Aciniform silk, which is used to wrap prey, and forms the outer lining of eggsacs, and, in males, is used to wrap up sperm for delivery during reproduction. Another use of aciniform silk by some orb web spiders is to decorate their webs with ornaments of various shapes, such as that seen in the webs of my PhD research subjects, the St Andrew's cross spiders, and other orb web spiders. (vii) Lastly, there is tubuliform silk, which is sometimes also called Cylindriform silk, and forms the inner coating of the spider's egg sacs, thus it is exclusively secreted by female spiders.

---

[*] Despite the name not all species build an orb-shaped web.

[†] Which includes iconic spiders like the Australian redback spider and the American black widow spider.

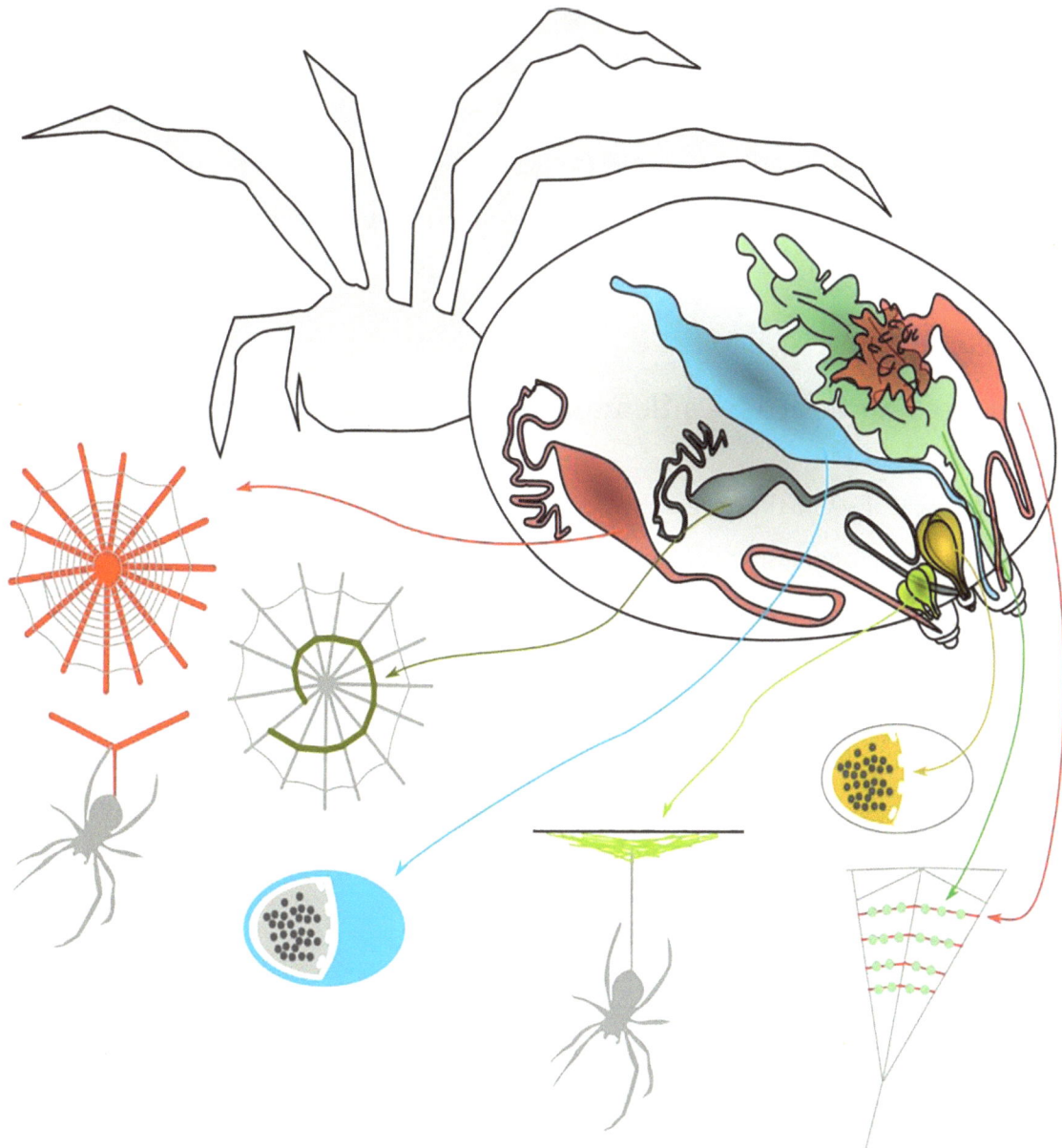

FIGURE 20. THE SEVEN TYPES OF SILKS PRODUCED BY ORB WEB SPIDERS AND THEIR GLANDS. FROM BLAMIRES ET AL. 2017.[2]

The glands secreting each of these silks bears the names of the silks they produce. For instance, major ampullate silk is produced by the major ampullate glands. Minor ampullate silk is produced by the minor ampullate glands. Aggregate silk is produced by the aggregate glands, and so on.

For each of the silks, as far as we know, a particular spidroin or set of spidroins* are secreted and stored by the glands as a concentrated solution, called dope. The dope flows through the gland wherein it thickens into a

_____

* While the various glands predominantly secrete a particular spidroin or set of spidroins (e.g. major ampullate glands specialize in major ampullate spidroins), it is not uncommon for a particular type of gland to secrete a range of different spidroins or for spidoins to migrate across glands through the spider's haemolymph. This means a particular type of silk from a particular gland might contain some spidroins different from those predominantly produced by the secreting gland.

gel before being drawn by the spider from a spinneret into the air (or water in the case of the diving bell spider *Argyroneta aquatica*). In spiders the spinnerets are just above the anus, or more correctly termed the anal pore, and not modified from the mouthparts as in moths. The dope immediately forms into a solid fibre upon exposure to the air. Solidification of the silk is promoted by the folding and aggregation of the spidroins during the final phase of spinning before it becomes fully dehydrated, whereupon the proteins lock in position as the silk is being extruded at the spinneret.

The spinning apparatus: a silk making factory within the spider abdomen

The silk glands are located within the spider's abdomen. In araneomorphs the silk glands take up most of the space in the front of the abdomen. Some of the glands are large (with a lot of variation across different spiders), like the major ampullate glands, and have regions that are distinctly identifiable to the naked eye upon dissection. The major ampullate gland, for instance, has distinct tail, sac (or sack), funnel, and duct regions (Figure 21). The tail region secretes the dope proteins and is followed by a sac that stores the dope at high concentration. The dope flows from the sac into the duct via the funnel prior to its extrusion at the spinnerets. Some of the other glands, for instance the pyriform and aciniform glands, are much smaller and lack any such easily discernible morphology.[2]

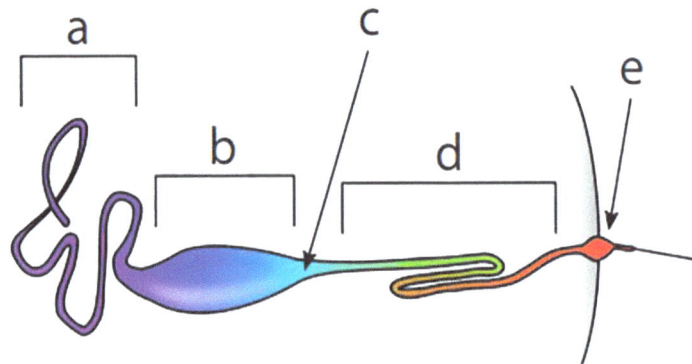

FIGURE 21. SKETCH OF THE MAJOR AMPULLATE GLAND WHERE (a) IS THE TAIL, (b) IS THE SAC, (c) IS THE AMPULLA, (d) IS THE DUCT AND (e) IS THE SPIGOT. THE COLOURS FROM PURPLE TO RED REPRESENT THE PH CHANGE, FROM pH8, WHICH IS SLIGHTLY ALKALINE TO A MILDLY ACIDIC pH5..

The silk dope flows through the glands by various means, depending on the gland in question, and is usually secreted as a solid fibre at the spinnerets. A given spider may have two or three pairs of spinnerets, with three commonly found in spiders of the Araneoidea. The three spinneret pairs are called the anterior, median, and posterior spinnerets based on their location; those closest to the base of the abdomen are the anterior, those closest to the spider's anus the posterior spinnerets, and those in between are the median spinnerets. In orb web spiders, the major ampullate and pyriform glands feed into the anterior spinnerets. The minor ampullate glands feed into the median spinnerets. The aciniform glands feed into both the median and posterior spinnerets, while both the aggregate and flagelliform glands feed into the posterior spinnerets. An evolutionarily very ancient* suborder of spiders from Southeast Asia, the Mesothelae, have a single type of silk and, weirdly, four pairs of spinnerets.

If we were to zoom in on any of the spinnerets using a high-powered electron microscope, we would see anywhere from two to over 50 spinning tubes called spigots associated with the spinnerets (Figure 22). In a particular group of web building spiders, the so called cribellate spiders, there are three pairs of spinnerets and a fourth sieve-like structure called a cribellum. While not considered a type of spinneret, the cribellum contains thousands of minute spigots, from which a particular type of fine, sticky silk, called cribellate silk, flows. This cribellate silk is used to retain prey caught in the webs of these spiders. A special comb-like structure called the calamistrum can be found on the forelegs of most cribellate spiders. This structure is used to 'comb' the cribellar silk as it is being pulled from the spinnerets to make them "woolly"[51]. I will delve into the structures and functions of the individual glands, spinnerets, and other structures in time, when I examine each of the different types of silk more closely.

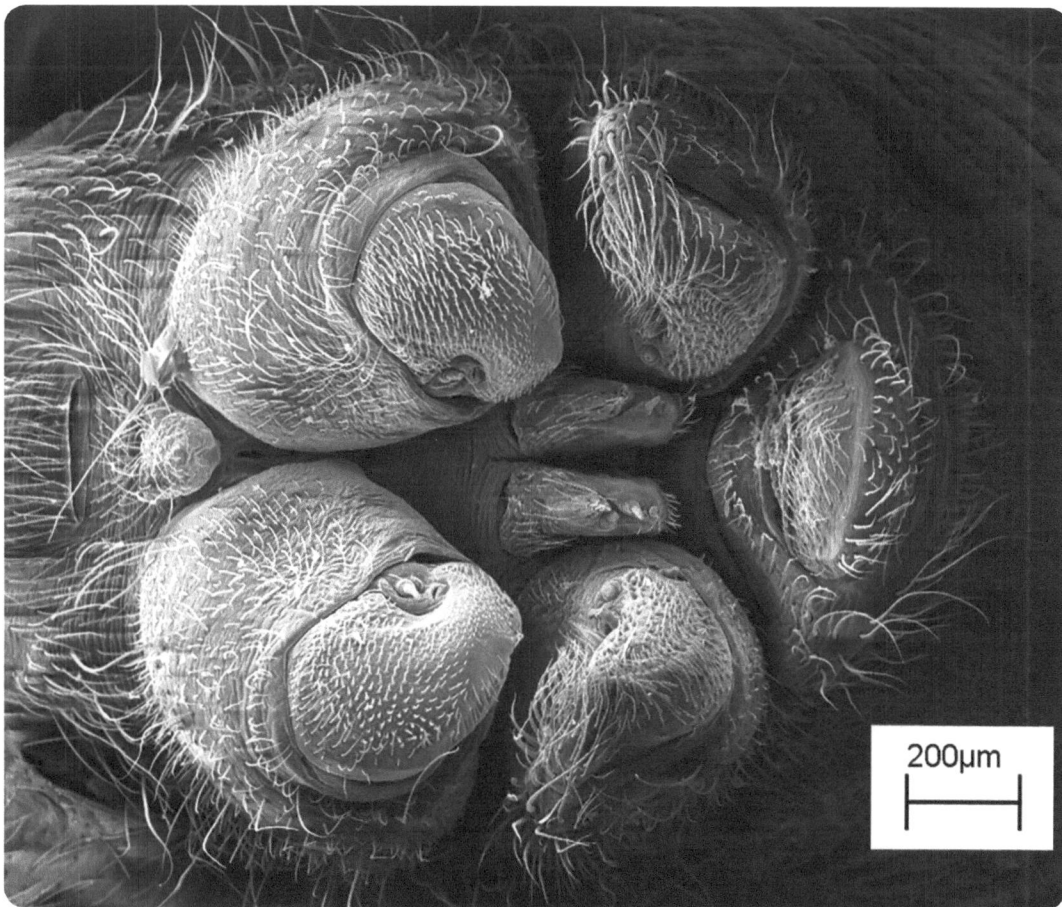

FIGURE 22. SCANNING ELECTRON MICROGRAPH OF SPINNERETS OF THE SPIDER *TRICHONEPHILA CLAVIPES*.

---

* Estimates date them as first appearing around 380 million years ago.

Major Ampullate silk: the heavy lifter

While most silks produced by spiders are quite tough, even in comparison to silkworm or bagworm silks, MA silk is exceptionally so. Mechanical testing of this silk has shown it to be one of the toughest materials going around. I even expect that comparative analyses will eventually reveal that it is much tougher than bagworm silk, but time will tell. Major ampullate silk has been estimated to be tougher than any man-made fibre, including a substance with the chemical name 1,4-phenylene-diamine and terephthaloyl chloride[52], that is commonly known as Kevlar®. That is, the fibre that bulletproof vests and other kinds of military protective clothing are made from. The extreme toughness of MA silk comes about because it has a unique combination of high strength and extensibility. Man-made materials are usually either very strong, like steel, or very extensible, like rubber but never both of these at once. MA silk is one exception to this generalization.

Within orb webs MA silk needs to be extremely tough because it is this silk, in the form of radial threads, that absorbs all of the energy of the insect smashing into it* when it strikes the spider's web.[2] It also absorbs all of the force applied by a spider when abseiling from its web onto the ground to escape danger, which may be from several metres high. Since orb web spiders often place their two-dimensional webs directly in the way of insect flight paths and the webs are intentionally made to be invisible to insects, the insects will often crash into the webs at full speed, which can exceed 50 kilometres per hour in some instances. At 2 to 5 micrometres in diameter, a dragline silk thread is miniscule compared to the insects they intercept. Scaled up, the feat of stopping an insect in full flight might be described as the equivalent to a 4mm thick rope or cable, or an average sized child's skipping rope, stopping a small to mid-sized passenger jet, like an Air Bus, in its tracks mid-flight. This is one of the myths about spider silk that I heard when I started working on spiders and it is, mathematically at least, quite factual for MA silk.

This remarkable silk comprises of a lipid and glycoprotein covered skin surrounding the fibrous protein core. The core contains at least two kinds of spidroins, called major ampullate spidroin 1 (or MaSp1) and major ampullate spidroin 2 (MaSp2).[2] These proteins are large, with a molecular weight of well over 300 kDa†, and have a highly repetitive central region flanked by non-repetitive N- and C- terminal domains. A silk gene sequencing project for the orb web spider *Araneus ventricosus* undertaken at Keio University in Japan revealed some possible additional MA silk spidroins. These included a major ampullate spidroin called MaSp3, which has a shorter repetitive region than either MaSp1 or MaSp2 as it lacks the poly-alanine repeating motif[53], and another called MaSp4.

The two best described major ampullate spidroins are MaSp1 and MaSp2, and these are thought to be coded for by two distinct genes, quite unimaginatively called *MaSp1* and *MaSp2*‡. The *MaSp1* and *MaSp2* genes have now been fully sequenced for the orb web spiders *Araneus ventricosus*[53], *Trichonephila clavipes,* and Darwin's bark

---

* Which is a product of both the mass of the flying insect and the velocity of its flight.

† 300 kilo-Daltons (or 3000 Daltons), with a Dalton being a unit of molecular mass that is standardized as the gram equivalent of one-twelfth the mass of a carbon 12 atom.

‡ By convention the gene is written in italics while the protein is written in plain text.

spider (*Caerostris darwini*). Darwin's Bark spider from Madagascar incidentally produces the world's strongest MA silk as well as the widest webs, which can span large rivers. A result of the accumulated knowledge of the major ampullate spider silk genes, their expression pattern, and their resultant physical properties, is that a database displaying full or partially sequenced genomes, and their correlating properties, has now been compiled by an international team of researchers for the silks of over 1000 spiders [55].

All the structure-function analyses done so far for spider MA silk suggests that the toughness of the silk's fibres is a consequence of the amino acid sequences of its spidroins and the secondary structures that these sequences subsequently form. The repetitive region of the MaSp1 protein consists of repeating polyalanine, (GA)n, (GGX)n and (A)n amino acid motifs (where G = glycine, A = alanine and X = other amino acids). These motifs form into cassettes that promote the formation of crystalline β–sheet nanostructures in the assembled fibres. The MaSp2 protein on the other hand has been thought to consist of multiple (GPGXX)n motifs (where P = proline), which predominantly form disordered β-turns and non-crystalline structures.[2] The combination of these structures is conducive to simultaneously providing high strength and extensibility; with the (GA)n, (GGX)n and (A)n motifs promoting fibre strength, and the (GPGXX)n motifs promoting its extensibility. A further consideration, revealed by a special analytical technique called Dynamic Nuclear Polarization (DNP)[57] is that the amino acid proline can, in some MA silks, attain a hydroxyl radical, which is a charged side group consisting of oxygen and hydrogen. The result of this is the formation of a substance called hydroxyproline within the (GPGXX)n motif. The presence of hydroxyproline within the (GPGXX)n motif changes the behaviour of the motif. It breaks hydrogen bonds along the protein chain causing them to slide easily over each other, resulting in slippage along the protein backbone, which manifests as an enhancement of the silk's extensibility.

Optimal toughness in MA silk is thought to be achieved when an approximately equal combination of MaSp1 and MaSp2 is found. Since MaSp2 is the only protein predicted to contain the (GPGXX)n motifs, proline composition has long been considered a reliable indicator of a given silk's relative composition of MaSp2[58]. One line of evidence for the idea that the ratio of MaSp1 and MaSp2 is responsible for its mechanical properties of spider MA silk lies in the finding that the more evolutionarily derived web building spiders can be distinguished from more ancient spiders by their silks having greater extensibility and toughness along with higher proportions of the MaSp2 protein[59]. Nevertheless, specific adaptations among the species that might influence silk amino acid composition and/or silk mechanics include them having inherently different diets[60]. This therefore cannot be ruled out as having additional influences over the known variations in MA silk properties.

Only by finding covariation in silk MaSp compositions and mechanical properties within and across species in a group of similar spiders from similar habitats would it definitively show that the expressed MaSp1 to MaSp2 ratio directly influences silk mechanics. To this end, my collaborators in Uruguay and I examined variability in the ratio of MaSp1 and MaSp2 and the subsequent mechanical properties in the MA silks of web building and non-web building wolf spiders from semi-rural environments in Uruguay.

Wolf spiders (that is spiders from the Family Lycosidae) belong to a spider group, or clade, within the Araneoidea called the Retrolateral Tibial Apophysis (RTA) clade; the name comes from a distinct projection on the male appendage (called a pedipalp) among these spiders. Most wolf spiders do not build webs. However, a select few species in the Subfamily Sosippinae build ground level, non-sticky, funnel shaped webs that capture insects that free fall into them. It might be expected that if the presence of MaSp2 is critical for silk toughness and this is necessary for webs to perform their function of capturing falling insects, then the silks of web building wolf spiders should be tougher, which should correspond with a greater MaSp2 composition, in web building wolf spiders.

We went ahead and tested this prediction by collecting MA silk from different species of web building and non-web building wolf spiders in rural Uruguay and tested their mechanical properties using a tensile tester housed at Deakin University's Institute for Frontier Materials in Geelong, Australia. We also examined their protein secondary structures using SAXS/WAXS, at the Australian synchrotron centre in Melbourne, and their amino acid composition using a procedure called High Performance Liquid Chromatography (HPLC). We found that the silks of the web builders were indeed much tougher. Moreover, the toughness of their silk corresponded with a greater amount of MaSp2, or at least a MaSp2-like protein, and this protein induced greater nano-structural alignment within the silk's amorphous region[61]. This study therefore shows rather definitively that the expressed MaSp1 to MaSp2 ratio influences silk mechanical properties in at least these spiders.

Plasticity: Spider silk's amazing changeability

One confounding problem with any suggestion that the ratio of MaSp1 to MaSp2, and perhaps other spidroins such as the recently described MaSp3 and MaSp4, are responsible for all of the variations in MA silk's mechanical properties is the frequent findings by different research teams that MA silk fibres often vary in property within a spider species, even within individual spiders, without any corresponding changes in the amino acid composition of the proteins[62].

In biology this kind of variability is called 'plasticity', or more precisely 'phenotypic plasticity', which is defined as there being measurable changes in some phenotypic trait, or traits, across environments without any corresponding change to the underlying genes or genotype. I have been reluctant to use the term plasticity in this context as in materials science the term plasticity describes a non-reversible change in a material's shape or function. To distinguish the terms when dealing with silks, I have instead used distinguishing terms like 'biological plasticity' and 'material plasticity'[2] or 'biological flexibility'.

The mechanisms inducing biological plasticity in MA silk may involve one, or a combination, of the following processes: (i) differential expressions of the MaSp1 or MaSp2 proteins*, (ii) physiological or biochemical processes

---

*   That is, the underlying genes, the genotype, is not variable but the way they are expressed is, which may include variations in factors or molecules (usually proteins) that regulate the transcription of DNA into RNA, or somehow modify and/or alter the DNA or RNA, or their associated histone proteins.

within the silk gland varying during spinning. Since factors such as sac or duct acidity or salt concentrations have effects on protein aggregation and folding, such factors can alter the physical properties of the spun fibres. Or, (iii) forces acting on the solid fibre during final drawing from the spinneret, or some kind of post-drawing change in silk property.

One experiment by researchers at the University of Oxford found that MA silk spun by spiders walking along a surface is not as stiff as that spun by free falling spiders[63]. In a different experiment, at the Universidad Politécnica in Madrid, Spain, SAXS/WAXS examinations of MA silk that had been forcibly collected from spider spinnerets with the speed of reeling manipulated showed that forces acting on the solidifying silk fibre during spinning affect the structural alignment of the proteins in the crystalline and amorphous regions, with subsequent influences on the fibre's mechanical properties[64].

I have done many studies, primarily with my colleagues in Taiwan, that show that MA silk properties are plastic across a range of environments. One example of such a study involved exposing spiders to winds of different strength. Our subsequent SAXS/WAXS experiments found that the spiders produced silks that differed in their crystalline and amorphous region structural alignment, leading to enhanced ultimate strength and stiffness when exposed to strong winds[65]. Another study showed that the giant orb web spider *Nephila pilipes* produces stronger and tougher MA silk when feeding on a high protein diet than it does when deprived of protein, and this is a consequence of spiders on the high protein intake producing silks with considerably denser crystalline β-sheet secondary structures[66].

My colleagues think, as do I, that wind exposure induces physiological or biochemical processes to change within the silk gland of the spider during spinning. It also induces variability in the forces acting on the silk during spinning, drawing, and the post-draw environment, each of which have been shown to affect the β-sheet alignment of the silk proteins. Nutritionally induced silk property plasticity on the other hand has been directly attributed to either the differential expression of MaSp1 and/or MaSp2 or physiological and biochemical processes acting during spinning. Studies have found that spiders on low protein diets produce MA silk with a lower composition of the amino acids proline and/or serine, probably as a consequence of lower MaSp2 production. Since proline and serine are only found in MaSp2 and are thought to be synthesized at a high metabolic cost, the costs involved with expressing MaSp2 compared to MaSp1 might, therefore, explain why some spiders vary the ratios of the spidroins they express in their silks when protein or energy intake is low.

Exactly what physiological and biochemical processes are acting during spinning and how they affect silk secondary structures are largely unknown by researchers. Anatomically the major ampullate gland consists of a tail, sac or ampule, funnel, duct and spigot, which opens at a valve. A pH gradient extends along the length of the gland, ranging from about pH 8 in the tail of the gland to about pH 5 at the valve, as depicted in Figure 21. Maintaining this gradient seems to be critical for the formation of the protein secondary structures[67]. MaSp1 and MaSp2 are secreted by the epithelial cells in the tail of the gland in various amounts depending on the species and amino acids accessed in the diet. The spidroins are then stored as a concentrated dope solution of anywhere from 20 to 50% protein in the region of the sac furthest away from the duct.

How it is that the proteins can be stored in such high concentrations without precipitating out is not well known, but hypotheses proposed including a suggestion that the proteins aggregate in the dope into circular structures called micelles[68]. Another hypothesis suggests they are condensed into a special hydrated polymeric substance called an aquamelt at high density with sodium, potassium, and calcium ions secreted by the sac epithelia to prevent the spidroins from aggregating out[69]. A proton pump and carbonic anhydrase enzymes maintain the pH of the sac at around 6.8-7.0, which is approximately the estimated isoelectric point, which is a measure of the pH at which all of the electrical charges of a molecule are neutralized, of MaSp 1 and MaSp 2[70].

The different chemical composition of the two spidroins ensures that they aggregate and fold differently from each other during the spinning process. As the dope flows from the sac into the duct across the funnel the width of the gland abruptly decreases, and the dope comes under an extremely high shear stress (which is a kind of rubbing stress). This induces a phenomenon called "shear thinning" of the dope. The duct length affects the amount of shear thinning experienced prior to spinning, the longer the duct the greater the thinning. The size and shape of the valve opening at the spigot varies across species and as a spider ages, explaining why adult spiders, with their longer spinning ducts than juvenile spiders, generally spin tougher silks than juvenile spiders[71`].

As the dope flows within the duct a reduction in the salt concentration induces polymerization* of the proteins by destabilizing the N-terminal. Phosphoric acid is released into the duct and this reduces its internal pH from about 6.8 to about 6.0, meaning the duct environment becomes more acidic. This acidity facilitates further protein aggregation, assembly, and polymerization into a fibre. The C-terminal domain acts as a 'switch' controlling the rate of the assembly[72]. The dope is now at extremely high concentration and resembles a semi-solid gel.[2] The large shear stress on the dope in this phase and an additional reduction in pH within the duct induce the proteins to continue folding to eventually form into β–sheets and other secondary structures. The rate of structure formation varies with dope concentration, glandular pH, temperature, and many other physiological and biochemical factors.

In the final stages of spinning the dope becomes increasingly gel-like and eventually pulls away from the wall of the duct, a phenomenon called draw down taper[73]. Under drawn down taper the proteins self-align and the silk dehydrates and solidifies into a fibre. Computer-derived atomistic modelling has predicted that additional frictional forces at the valve might cause the crystalline and amorphous chains to align differently[74]. Spiders may be able to change the frictional force at the valve to alter the alignment of the secondary structures and "tune" the performance of the silk fibre. This explains why it is that the silk spun by spiders in free fall are thicker and more compliant compared to those of spiders crawling on a horizontal surface, and the silks produced by spiders that had been anaesthetized prior to silk collection are thinner and stiffer than silks from spiders that had not been anaesthetized.

---

* Polymerization is the chemical joining of the individual polymers, in the case of silk their constituent proteins.

When laboratory spun MA silk is drawn into different media, including air, water, urea, or ethanol, during artificial spinning its properties differ substantially. MA silks spun in air for example are stiffer and stronger than those spun in water, urea or ethanol under similar tension because exposure to dry air completely dehydrates the silk fibres enabling hydrogen bonds between proteins chains to reform and the fibre to stiffen[75]. Silks spun into ethanol have similar properties to those spun in water because both water and ethanol are polar, that is they have some positive and negative charges at a molecular level, and so are effective at disrupting hydrogen bonds between protein chains in the amorphous region, rendering the silk highly elastic.

MA silk spun in water, or that spun in air but later immersed in water or almost 100% relative humidity upon spinning, are extremely soft, extensible, and tough. This phenomenon of water affecting the property of spider silk post-spinning is known as "supercontraction" and is reversible upon removal from water or a reduction in humidity.

As the name implies the silk shrinks or contracts, sometimes to over half its original length. Supercontraction occurs because water molecules disrupt the hydrogen bonds between the crystalline β-sheets, mobilizing the proteins thus inducing significant misalignment of the proteins in the amorphous region. This misalignment of proteins induces overall fibre shrinkage, which may be up to 50% depending on the spider and the amino acid composition of the silk, as it enables the silk proteins to become mobile and easily slide past each other. Since supercontraction removes the influence of any final spinning effects on protein alignment, as well as the overall size and alignment of the crystalline and amorphous region components and, ultimately, the silk's mechanical properties, the supercontracted state may be considered as a "ground state" for MA silk. Supercontraction can thus be utilized to return the tertiary structures of MA silks from different spider species to a state that is independent of any shear or drawdown taper effects. It is accordingly useful for comparing the raw properties of MA silk from phylogenetically distinct spiders[77].

Researchers at the Universidad Politécnica de Madrid in Spain have gone a step further and derived a constant, called the α^* (or 'alpha star') which they calculated from the supercontraction shrink that any given spider silk experiences upon exposure to water. The formulation of the constant has allowed researchers to compare the mechanical properties of major ampullate (MA) silks of different spiders in a standardized way. It was used for example to assess silk property variation of different spiders from the same urban habitat to help researchers focus in on the evolutionary drivers of spider silk property variation[78].

When a spider spins a silk fibre in air, dehydration causes the secondary structures of MA silk protein chains to lock in position. Nevertheless, silk fibres can still change in property within air. In dry or hot environments, MA silk becomes stiff and eventually brittle as a consequence of molecular vibrations displacing hydrogen bonds between the proteins causing slippage and eventual disruption of the connections between protein chains[79]. When exposed to direct sunlight MA silk, and other silks within spider webs or cocoons, experience short and/ or long-term decreases in strength and elasticity as a consequence of UV and heat damage to the skin and/or core spidroins[80]. The proposed ecological and evolutionary significance of plasticity in MA silk is that it enables web building spiders to tune the properties of their silk fibres to perform differently as their needs change across ecological or behavioural contexts. From an engineering point of view, it shows that a single set of building blocks can produce differently performing materials when in different environments and provides for a natural starting point for building environmentally adaptable "smart materials" for a range of practical purposes.

Minor Ampullate silk: Major's flexible friend

Minor ampullate (MiA) silk has a lot in common with major ampullate silk. Like MA silk MiA silk is secreted as a highly concentrated proteinaceous dope by a gland called the minor ampullate gland. Two minor ampullate silk proteins have been described to date, minor ampullate spidroin 1, or MiSp1, and minor ampullate spidroin 2, or MiSp2. Like major ampullate spidroins these spidroins have a repetitive region and non-repetitive N- and C- terminal domains. These proteins are coded for by between four 7-9kb long *MiSp* genes that have been called *MiSp*a-*MiSp*d[54] by some and *MiSp1a, MiSp1b* and *MiSp2a, MiSp2b* by others[81].

The amino acid compositions of the repetitive region and N- and C-terminal domains of MiSp1 and MiSp2 does differ from those of MaSp1 and MaSp2. While both proteins contain similar (GA)n, (GGX)n and (A)n amino acid motifs, with (GA)n, and (GGX)n most predominant, the repetitive regions of MiSp1 and MiSp2 are shorter than those of MaSp1 and MaSp2 (at 1700 amino acids long) and neither contains (GPGXX)n motifs. Instead, different motifs such as GGXGGY (Y being other amino acids that are not the identified X amino acid), (GA)n, and GAGA appear in different quantities in both proteins. The physical and chemical properties of the minor ampullate silk proteins, accordingly, differ significantly to those of MaSp1 and MaSp2, and so provide minor ampullate silk with distinctly different mechanical properties to those of MA silk.

Comparative studies of different species of orb web spiders have found that minor ampullate silks do not differ as much between species and among individuals as major ampullate silk. Compared to MA silk, MiA silk exhibits lower modulus and tensile strength but slightly greater extensibility[82]. Nevertheless, minor ampullate silk does not supercontract in water[83]. In this respect, minor ampullate silk might better resemble silkworm silk, and most other types of silk for that matter, than MA silk. This is most intriguing as the size, density and alignment of the crystalline nanostructures are similar to that found in major ampullate silk yet it does not supercontract.

The minor ampullate gland resembles the major ampullate gland in shape and its anatomical components but is considerably smaller in size; hence its name. It has a tail, albeit a smaller and shorter one than the major ampullate gland, an enlarged sac in which the spidroins are stored as a concentrated dope, and a duct through which the dope flows prior to its extrusion as a fibre. Genetic evidence nonetheless suggests that the minor ampullate gland is more likely to be evolutionarily derived from the tubuliform or aciniform gland than the major ampullate gland[84]. The similarities in shape of the major and minor ampullate gland accordingly seem to be an example of convergent evolution in silk gland morphology.

Flagelliform silk: The most elastic silk

Flagelliform silk forms the underlying fibre in the capture spiral threads in spider orb webs. It functions to capture insects in flight and to prevent those caught from escaping. Genomic analysis has revealed that the single flagelliform silk protein Flag, or AvFlag (Av being for the spider from which the protein was first isolated; *Araneus ventricosus*) consists of a 167 amino acids, with a repetitive region and an 87 amino acid C-terminal non-repetitive region. It appears to lack an N-terminus[85].

The primary and secondary structures of the flagelliform silk fibre have been ascertained for the orb web spiders *Araneus ventricosus* and *Trichonephila clavipes* by procedures such as SAXS/WAXS, solid-state Nuclear Magnetic Resonance (ssNMR) and others. The studies showed flagelliform silk to comprise of a similar lipid and glycoprotein rich skin and proteinaceous core structure as MA silk[86]. The core of the fibre has long been thought to just comprise of Flag proteins. However, the recent mapping of the *Trichonephila clavipes* silk genome[54] has described a second Flag protein. The authors of the subsequent report named the original Flag as Flag-a with the new protein designated as Flag-b. More information on the amino acid sequences and characteristics of Flag-b is needed to ascertain whether it is really a new protein, however.

A SAXS/WAXS analysis of flagelliform silk found that, in contrast to major and minor ampullate spidroins, Flag lacks the polyalanine motif, thus it forms very few crystalline β-sheet secondary structures. Accordingly, the fibres have around half the tensile strength of MA or MiA silk. The repetitive region of flagelliform proteins, however, comprises of about 75% GGX, GPGGX, and other highly conserved repetitive motifs that function to induce a nano-spring-like functionality to parts of the fibre. The high abundance of these motifs in flagelliform silk provides it with immense extensibility, a bit like a rubber spring. Indeed, with an extensibility seven times greater than that of MA silk, flagelliform is by far the most extensible silk produced by spiders. If you have ever watched an insect or any other object strike an orb web at full flight, you can see this extensibility in action when the spiral threads of the web extend well beyond their original length to cushion the impact of the object striking the web. Interestingly, a recent proteomics analysis (a complete study of the protein structure and function) of the Flag proteins, revealed the presence of much hydroxylated proline in flagelliform silk[87], lending further credence to the idea that hydroxylated proline, or hydroxyproline, is required for high extensibility in silks.

While extended, the capture threads of an orb web behave as a rubbery solid, owing to the nano-spring-like behaviour of the GGX, GPGGX, and spacer motifs. Upon relaxation, however, they switch behaviour to function more like a viscous liquid. The flagelliform thread becomes partly adsorbed within the aggregate droplets. The subsequent shortening of the thread returns the web to its original tension, superficially functioning in a somewhat similar manner to a windlass. The behaviour of flagelliform silk thus provides us with a valuable blueprint for biologically inspired designs of functional solid-liquid hybrids, and highly elastic devices created therefrom.

Aggregate silk: It glues, grips and holds tight

Aggregate silk is a sticky viscous silk that coats the flagelliform capture threads of spider orb webs. It is the stuff that entangles you completely and is seemingly impossible to remove from yourself if you walk through a spider web. A type of aggregate silk, with somewhat different adhesive and chemical properties, is found covering the bottom centimetre or so of the so-called gumfoot threads of cobwebs.

Cobwebs are the haphazardly shaped three-dimensional webs built by spiders such as black widows and redback spiders. They comprise of at least three silken components: (i) a retreat where the spider positions itself, which sits behind the (ii) sheet threads through which the spider moves, and (iii) the gumfoot threads which run from the sheet to the ground. At the base of these threads lies the aggregate silk-derived gumfoot glue (see Figure 23). The gumfoot threads, incidentally, have an axial core of major ampullate silk rather than flagelliform silk. The primary function of aggregate glues in orb webs and cobwebs is to combine with the underlying silk thread, whether it be flagelliform silk in the case of orb webs or major ampullate silk in the case of cobwebs and retain the captured insects.

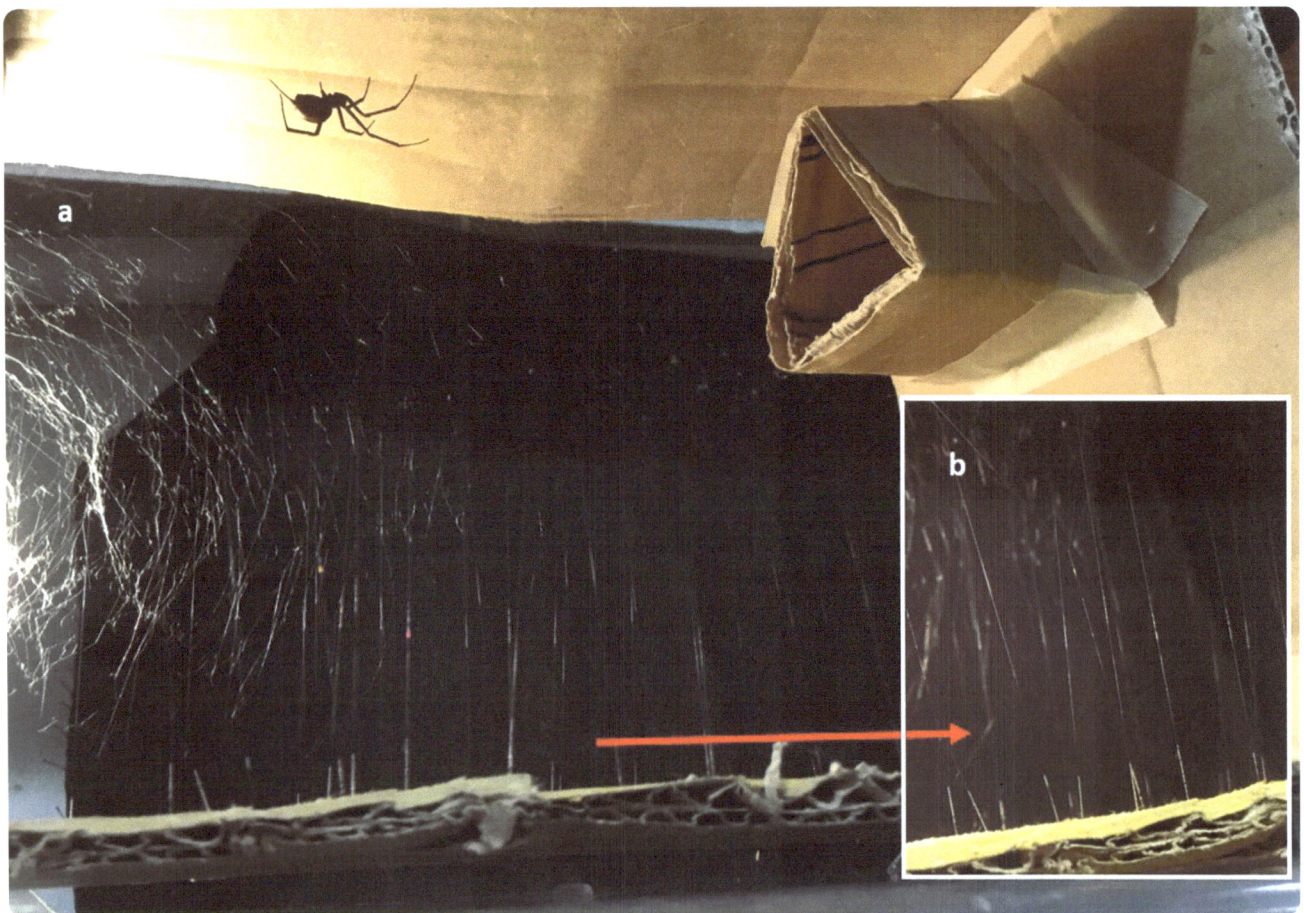

FIGURE 23. WEB OF THE REDBACK SPIDER, WITH A CLOSE UP OF THE GUMFOOT THREADS SHOWING THEIR ENDS COVERED WITH AGGREGATE GLUE.

In orb webs, the flagelliform axial thread of the capture spirals deforms as the insect struggles to break free so as to give it no traction to push itself out of the web. The liquid aggregate silk then functions as a glue and adheres strongly to the cuticle of the captured insect. This gluey substance contains a mixture of proteins, glycoproteins and low molecular weight organic (including the neuropeptides GABamide, choline, isethionic acid) and inorganic (including hydrogen sulfate and potassium nitrate) charged compounds, which have been collectively referred to as the silk's organic and inorganic salts*, primarily because they are dissolved water[88]. The glue's stickiness is conferred by the both the glycoproteins and salts joining forces to promote the generation of strong cohesive forces† within the solution. These forces also cause the aggregate silk to coalesce into droplets along the axial thread. Incidentally, this process is called Plateau-Rayleigh instability, and it is what induces water to coalesce into droplets as it falls from a tap or as rain.

The salts promote the uptake of additional water from the environment, which helps keep the glycoproteins hydrated so that they remain sticky and extensible. This is important as it is the elasticity of the glycoproteins upon binding to insect cuticle that prevents the insects from escaping the web.[89]. It also keeps the flagelliform thread in a state of supercontraction, thus ensuring it retains its compliance and recoverability which means it is not easily displaced when prey impact the web, or when the prey struggle to escape. By doing experiments where investigators have peeled orb web capture threads off from a smooth surface, such as glass or stainless steel, it has been found that multiple glycoprotein units combine and adhere to a captured prey item all at once. The adhesive has been shown to function in webs as follows. The glue first encompasses the trapped insect in sticky threads. Then, as the insect moves in an attempt to struggle free from its entanglement the compliant axial threads stretch along the length of the flagelliform thread. Magnification of the spiral thread fibres of an orb web being pulled off a flat surface have revealed that the stretching glue droplets resemble the ropes of a suspension bridge (Figure 24). The mechanism by which a gluey thread adheres and flexibly moves when stuck to a surface, such as an insect's cuticle, has accordingly been called a "suspension bridge mechanism"[90].

---

* These differ from conventional salts which are solid lattice whose charged particles are held together by strong ionic forces (i.e. electrical charges).

† Cohesive forces are the forces that make molecules become attracted to themselves. This compares with adhesive forces which make dissimilar molecules attracted to each other.

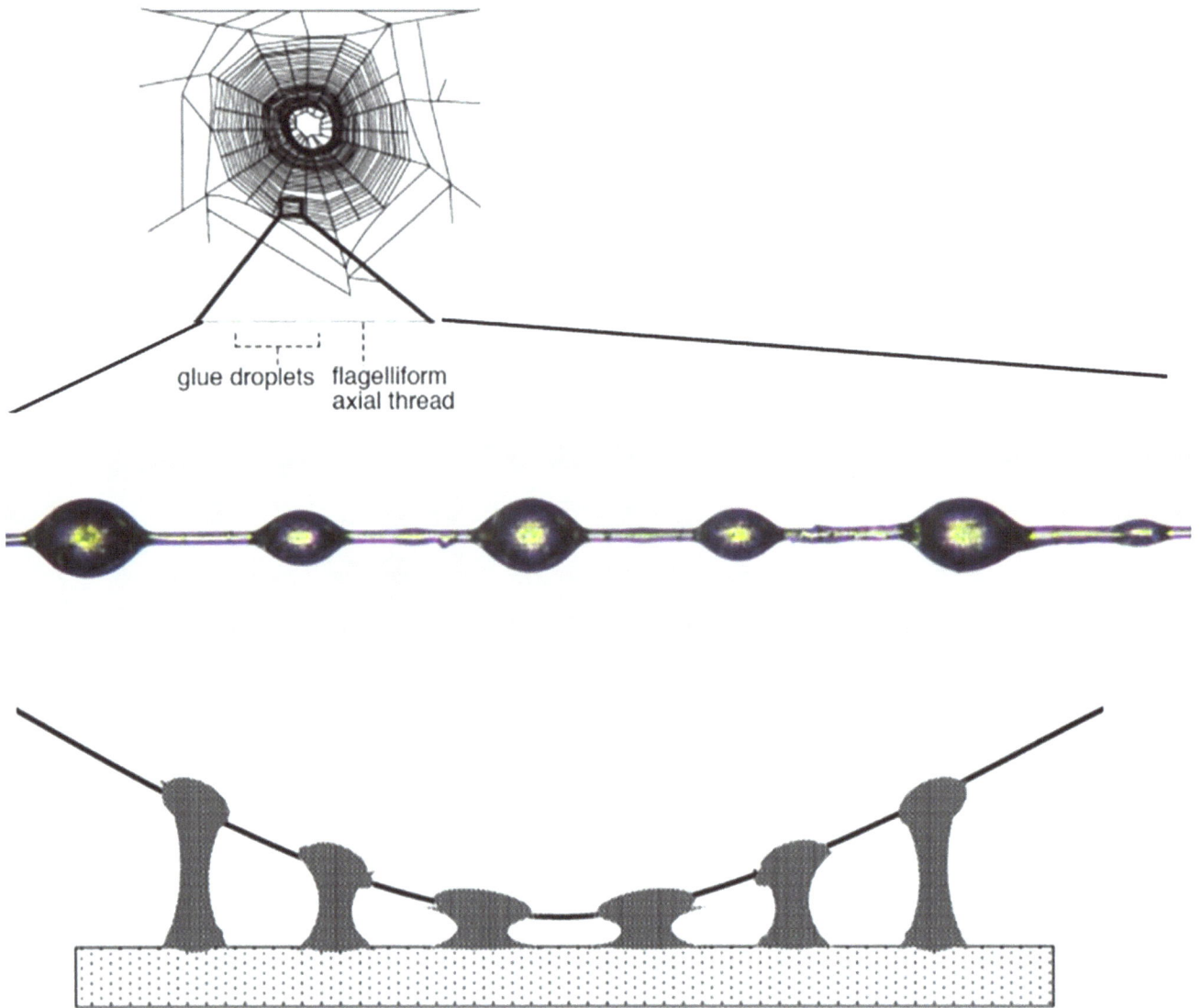

glue droplets   flagelliform
axial thread

FIGURE 24. THE SUSPENSION BRIDGE MECHANISM OF ADHESION BY THE AGGREGATE GLUE DROPLETS OF ORB WEB SPIRAL THREADS. FROM BLAMIRES 2019[91].

The high hygroscopicity, i.e. its ability to readily absorb atmospheric moisture, of the aggregate droplets means that they swell in humid environments, but have a tendency to shrink in dry environments. The degree of swelling and shrinkage across environments differs among spider species, probably as a result of the different mixes of salts found in their aggregate glue droplets. To some extent the mix of salts in the aggregate droplets of any given species is adapted to the environment the spider lives in, meaning that the droplets of spiders from humid environments lose water if high humidity is not sustained while those from drier environments can retain water even if humidity drops severely[92]. There are nonetheless limits to this and most orb web aggregate spider's glues will cease to function in extremely dry conditions.

Experiments have found that orb web aggregate glue droplets shrink or swell as a result of an individual spider's diet, most likely because differences in the dietary uptake of nutrients and salts induces there to be differences in the mix of salts within the spider's aggregate glues.[93] The aggregate glue droplets that form on cobweb capture threads nonetheless do not swell and shrink with humidity or dietary changes to the same extent as those in orb webs[94]. It is unclear why there is such a disparity in the behaviours of orb web and cobweb spider aggregate silks, but it certainly points to there being significant differences in the mix of the constituent salts.

One consequence of the variability in glue droplet volume across environments or diets is a concurrent variability in glue stickiness, with the larger droplets often making the spirals stickier. It has, for instance, been found that spiders that had been feeding exclusively on cockroaches produce larger and more sticky droplets than those feeding exclusively on crickets[95]. There could be some kind of ecological advantages for such diet-induced changes in stickiness. One experiment, for example, found that the webs of spiders starved of protein retained their cricket prey longer than those fed high protein feeds[96]. We accordingly think that spiders are somehow able to tweak the compositions of the salts in their aggregate glues to retain some prey for longer to provide better opportunities for the spiders to attack and wrap the prey when under starvation conditions. It could be that the spiders will be happy to invest in sub-optimally performing spiral glues when well fed because the salts are metabolically costly to synthesize[97].

How and why the orb web spiral threads become stickier when the aggregate glue droplets expand in volume is not known. There seems to be a limit to the increase in stickiness as the droplets expand. Indeed, experiments exposing orb web spirals to increasingly higher humidity show them to eventually become less sticky if the droplets expand beyond a certain size, a phenomenon attributed to over lubrication of the adhesive glycoproteins in the droplet's core[98].

Aggregate silk is secreted by the relatively small aggregate silk gland, which is only found in orb web building spiders and their relatives, including the cobweb spiders. Whether aggregate silk was within the toolkit of ancestors of orb web building spiders is not known, but some molecular evidence suggests it might have been[99]. Analyses of the genes expressed by the cells of the aggregate gland of some orb web spiders have revealed a couple of unique spidroins, called ASg1 and ASg2, as well as other proteins and gene products unique to the gland[100]. The gland and associated genes accordingly seem to have evolved more recently than the other spider silk glands, including the closely associated flagelliform glands. Aggregate spidroins are secreted into the aggregate gland duct where they are glycosylated prior to secretion along with flagelliform silk at the posterior spinnerets. Flagelliform silk is secreted by a single spigot with two adjoining spigots secreting the aggregate silk that it encovers. These flagelliform and aggregate secreting spigots have been referred to as the flagelliform-aggregate "spigot triad" (see Figure 25).

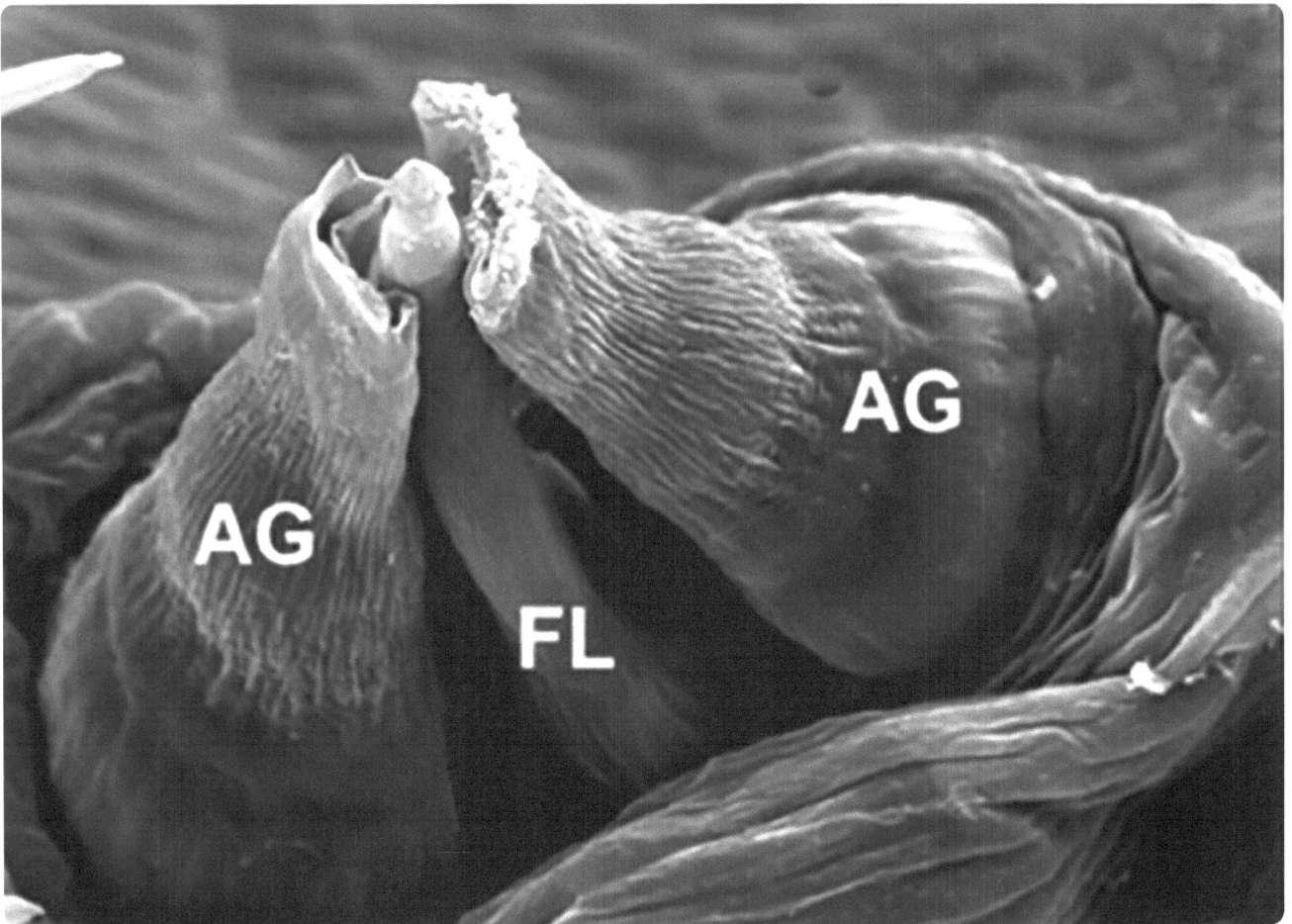

FIGURE 25. MICROGRAPH OF THE AGGREGATE (AG) AND FLAGELLIFORM (FL) SPIGOTS OF SPIDERS, WHICH FORM THE SO-CALLED SPIGOT TRIAD WHICH SILK SPINS THE GLUEY AGGREGATE AND FIBROUS FLAGELLIFORM SILKS.

Pyriform silk: Nature's anchors

Pyriform silk is a cement-like silk secretion. It lacks a well-defined secondary structure and emanates from the small grape-shaped pyriform glands which open at the anterior spinnerets. The major function of pyriform silk is to form the cementing attachment discs that anchor other silk, usually MA silk, to some kind of surface or another silk. As they must provide for robust and stable anchors for webs, draglines and egg sacs, stabilize the long-distance jumps performed by jumping spiders, and enable movement on, across, and within water or moist surfaces, the performance of pyriform silk must be able to be tuned to within very precise limits.

Various studies have quantified the adhesive properties of pyriform silk by measuring the forces required to pull the attachment discs off a substrate. These have found there to be substantial variability in attachment strength depending on the species or mass of the individual spider secreting them, or the substrate onto which the attachments are made[101]. The performance of the silk itself is difficult to measure, primarily because the pyriform attachment is difficult to isolate from the embedded MA or other silk. Nevertheless, it seems that the chemical composition of the silk is uniform across species and individuals and the only way spiders can mitigate any changes in attachment disc properties is via changes to the architecture of the attachment disc it forms, and this is only achieved by the spiders changing the way they sweep their pyriform producing spinnerets across the substrate. One study described two types of attachment disc architectures being utilized by black widow spiders. One being what was called a dendritic architecture, which is characterized by the spinning of elongated finger-like projections onto the substrate with little cement covering the underlying threads, and the other a staple-pin architecture, characterized by a tightly wound bundle of parallel threads embedded deeply within pyriform cement. Moreover, high speed video recordings of the spinnerets in action have revealed that the two architectures do involve different types of spinneret movements to generate.[102]

Regardless of the attachment disc architecture, pyriform attachments have a unique inherent combination of extreme internal flexibility and external rigidity. Microscopic techniques, such as freeze fracture and scanning electron microscopy, of the attachment discs have revealed the cement-like adhesive discs contain a nanofibrillar structure that provides for the internal flexibility, while embedded supporting lipids provide the means for its external rigidity[103].

Studies using genetic and proteomic methods have shown that the pyriform attachment discs of black widow spiders primarily comprise of a spidroin called PySp1[104]. A second proline rich protein, PySp2, has been proposed based on a full protein analysis of the attachment discs of the orb web spider *Trichonephila clavipes*[105]. A comparison of the full sequences of the *PySp1* gene from black widows and the distantly related orb web spider *Argiope argentata* revealed that there are many similarities in the PySp proteins of different spider species[106]. These studies have revealed that PySp1 consists of five distinct regions: (1) a 146 amino acid N-terminal domain; (2) a 498 amino acid N-terminal linker region; (3) a 4935 amino acid repetitive region (although this region comprises over 85% of the protein it is still substantially shorter than that of all the other spidroins); (4) a 90 amino acid C-terminal linker region; and (5) a 90 amino acid C-terminal domain. A full translation of black widow PySp1 found it to contain the highest alanine composition among any of the spidroins at over 60%. Surprisingly, however, little, if any, glycine (>5%) has been attributed to it. It also has a higher glutamine composition (~15%) than any other spidroin[105]. The amino acid composition of the residues making up PySp1 suggest it is a hydrophilic silk, making it soluble in water, and that the proteins are stored in and secreted from the pyriform gland at an exceptionally high concentration. Studying the physical and biochemical mechanisms by which pyriform silk is stored and excreted might be useful in assisting the development of systems that use highly soluble proteins or other polymers for spinning high performance adhesives or other materials.

Aciniform silk: For wrapping and decorating

Aciniform silk is used in a variety of ways depending on the spider. These include wrapping up prey prior to consumption, as an inner lining of eggsacs, by males to wrap up sperm before transferal to a female, and for decorating webs with unique silken shapes and ornaments called decorations or stabilimenta, as my PhD subjects did. The aciniform gland is small with no visible sac or tail compartments. The silk is not as strong as major ampullate silk, but is slightly more extensible and about as tough, which is impressive given the gland's morphology is quite simplistic compared to the major ampullate gland.

As with other silks, its properties are mostly a consequence of its primary and secondary protein structures. A spidroin called AcSp1 has been isolated from the aciniform gland of black widow spiders. Sequencing of the *AcSp1* gene has revealed that, like ampullate silk spidroins (that is MaSp1, MaSp2, MiSp1, and MiSp2), AcSp1 has a long repetitive region consisting predominantly of domains of repeating alanine and glycine. However, AcSp1 is virtually devoid of any proline. There is evidence from the *AcSp1* genetic sequence to suggest that AcSp1 might be an evolutionary precursor of the MaSp proteins[107]. For one, the compositions of the N- and C-terminal regions in the aciniform silk are similar to those in major and minor ampullate silks, and flagelliform silk. Also, solid-state NMR analyses of freshly spun aciniform fibres have shown that the spidroins are well aligned along the fibre backbone and comprise over 50% alanine and glycine rich crystalline β-sheets, just like major ampullate silk fibres. These structures are also similarly responsible for the silk's strength. Additional (GGX)n units and the occasional (GPGXX)n units form type II β-turns and resemble those found in insect elastin and are thought to contribute to the silk's high extensibility[108].

Aciniform silk differs from all the other spider silks in a unique way, that being it is highly reflective of UV light.[109] Spiders may exploit this property of aciniform silk when they add decorations to their webs. Indeed, it is widely thought that spiders decorate their webs to provide some kind of signal to insect and/or bird predators or prey[110]. The property of reflecting UV makes the decorations highly attractive to many insects, so it seems that a prey attracting mechanism seems quite likely. Furthermore, this hypothesis has been backed up by a number of independent studies. Nevertheless, there seems to be some experimental evidence for the use of aciniform silk decorations within webs as protection. Regardless of whether they are used for prey attraction, predator avoidance, or even parasite avoidance. There is great variation in their use and forms among different spider species and across ecological and other circumstances. Colleagues of mine at Tunghai University in Taiwan, for instance, found that different shaped decorations give different signals to different prey, predators, and parasites, so a combination of these mechanisms may drive the evolution of decoration use among certain spiders[111].

The type and shape of the silk decoration used depends mostly on the species of spider. Orb web spiders of the genus *Argiope* are the most prominent users of decorations, with almost all of the 70 or so known species adding some kind of visible silk decoration to their web. One of the most abundant and well-known examples is the cross or X-shaped decoration (the so called 'cruciate decoration'). Many spiders of the genus *Argiope* use these kinds of decorations, including *Argiope keyserlingi*; the species I studied for my PhD thesis. These spiders appropriately go by the common name of St Andrew's cross spiders or cross spiders. Other species of *Argiope* add a decoration of a single woven silk line running vertically down the web (or a 'linear decoration'), while others build a circular mass of silk around the hub of the web (a 'circular decoration'). There are even some species and/or individuals who use a bit of each type of decoration at once or change the type it uses as it ages. Sometimes spiders might change up the type or shape of the decoration they use from day-to-day (as depicted in Figure 28). In another study by my colleagues in Taiwan it was found that if the same decoration type is used repeatedly by an individual spider too many times its prey and predators learn to associate a particular signal with the presence of a spider, so the likelihood of catching prey diminishes over time while the likelihood of being eaten by a predator increases. These findings suggest that it is important for individual spiders to change their decoration patterns frequently[112].

FIGURE 26. THE DIFFERENT FORMS OF WEBS DECORATION SPUN BY THE SPIDER *ARGIOPE AEMULA* IN TAIWAN. THE SPIDER IN **a** HAS FULLY CRUCIATE DECORATIONS, **b** AND **c** ARE PARTLY CRUCIATE AND **d** HAS NO DECORATIONS.

Cyclosa is another genus of orb web spider containing many species known to decorate their webs. These are very small white, grey or mottled coloured orb web spiders found predominantly in Europe and North America, but some species occur in eastern Asia and Australia. In addition to circles and lines, these spiders can make very elaborate decorations, including wide and flat ovoid-shaped decorations, or decorations that look like swirls, irregular circular decorations that look like bird poo when the spider sits in the middle of it, owing to the spiders grey body (Figure 27),[113] and some even make decorations that look somewhat like a giant spider. Another example of silken web decorations, presumably made from aciniform silk, includes the "tuft" decorations used by a type of Asian spiny spider to decorate their barrier webs.* Unlike most other types of decorations these are thought to uniquely signal to birds to warn them not to approach the invisible but impenetrable barrier web that the spider has surrounded its orb web with[114].

FIGURE 27. THE SPIDER *CYCLOSA GINNAGA* SITTING ON ITS WEB CONTRASTING ITS BODY AGAINST ITS DECORATIONS. IT SEEMS LIKLE IT DOES THIS EXPLICITLY TO LOOK LIKE A BIRD DROPPING.

Tubuliform silk: A mother spider's means of protection

Most spiders coat their eggs with a layer at least 5 μm thick of tubuliform silk, around which they build the protective cocoon using aciniform or major or minor ampullate silk. The gland is long and tubular (hence its name), and lacks a distinct region for storing secreted proteins[115]. A spidroin called tubuliform spidroin 1 (TuSp1)

---

* Barrier webs are a common additional structures placed around the webs to act as a barrier to predators that might attack the spider.

has been identified in the tubuliform silk of orb web spiders[116]. Interestingly, a protein with a TuSp1-like sequence has been isolated from honeybee silk, but it seems unlikely that this protein is in any way evolutionarily related to TuSp1. It just happens to have some similar amino acid sequences.

TuSp1 primarily forms helical secondary structures in solution within the spinning duct, but its high alanine and glycine composition (over 50%) promotes the development of a range of crystalline β-sheet and β-turn structures upon its spinning in air. The size, distribution and stacking arrangement of these crystalline structures influences the strength and extensibility of the tubulifom fibres, each of which has been estimated to be almost as high as major ampullate silk, despite TuSp1 being a much smaller protein (~100 kDa) than the MaSp proteins[117]. Its high toughness, combined with a unique cross-weaving patterning of the silk upon its deposition, promote interlocking and piling* of the threads and make the tubuliform layer around the spider eggs exceptionally robust[118]. As alluded to when discussing the properties of moth silks, it may be that extreme toughness and robustness is a necessity in cocoon silks in order to provide extra protection for the eggs, as they may be exposed to highly variable environmental conditions and predators.

An additional protein called egg case proteins (ECP1) with a molecular weight of about 100 kDa has been isolated from black widow spider cocoons. ECP1 is secreted by the tubuliform gland and associates closely with the tubuliform fibres. It is somewhat similar to TuSp1 in composition with poly alanine and poly alanine-glycine repeating sequences, although the length of these repeating sequences is shorter than in TuSp1. The N-terminal domain of ECP1 differs from that of TuSp1 as it promotes the formation of disulfide bonds between proteins, and these prevent the proteins from slipping past each other. Accordingly, ECP1 has extremely high stiffness within the egg sac. ECP1 is not sticky so does not appear to bind the eggsac fibres in a similar way as sericin does in silkworm cocoons[119], rather it is the interlocking of the tubuliform silk threads that holds the cocoon together.

Unlike other spider silks, tubuliform silk can be a range of different colours, including white, yellow, grey, black, green, and purple (Figure 28)[120]. Studies of silk colouration[13] suggest that silk secondary structures do not affect the colour of silks, but it is the presence of pigments such as phenols, porphyrins, quinones and carotenoids that appear to be responsible for non-white silk colours. It would be interesting to examine the range of pigments expressed in the tubuliform and aciniform silks of different spiders to see whether there is much variability in their expression within and between species.

---

* Piling in this context is a repeated looping and warping pattern within the silk threads.

FIGURE 28. EXAMPLE OF THE DIFFERENT COLOURS OF EGGSACS OF SPIDERS.

Cribellate silk: Ancient twists

Cribellate spiders are an evolutionarily ancient group of orb web spiders that are specifically characterized by the presence of a silk spinning organ located on the lower abdomen that resembles a spinneret called the cribellum. It specializes in the spinning of cribellate silk, a unique type of silk in horizontally aligned orb webs, and is thought to have been derived from a second pair of anterior spinnerets that are still found in the most ancestral group of living spiders, the Mesothele and Mygalomorph spiders. Cribellate silk is spun as a puffy or woolly silk composed of numerous (numbering anywhere from in the 10s to in the 1000s) of individual nano-fibres that are extremely thin (about 10-30μm in width). The nanofibres are, as shown in Figure 29, surrounded by a few larger axial fibres which are of unknown glandular origin. The cribellum comprises of 1000s of individual spigots and the fibres are thought to be spun under high pressure using electrostatic forces, which appear to be facilitated by the rapid movement of the spider's forelegs, as a viscous fluid that solidifies on contact with air[121]. While it is not known whether or not the spigots of the cribellelum are fed by thousands of individual glands, it does seem unlikely. Due to their size, the structures and properties of cribellate threads are largely unknown, but analyses of the bulk threads have enabled some generalizations across species and families to be made. The broad microstructure of the cribellate threads of the small uloborid spider *Uloborus plumipes*, have been probed by scanning electron microscopy and atomic force microscopy, and it appears that the fibres comprise of highly segmented α-helical protein secondary structures[122].

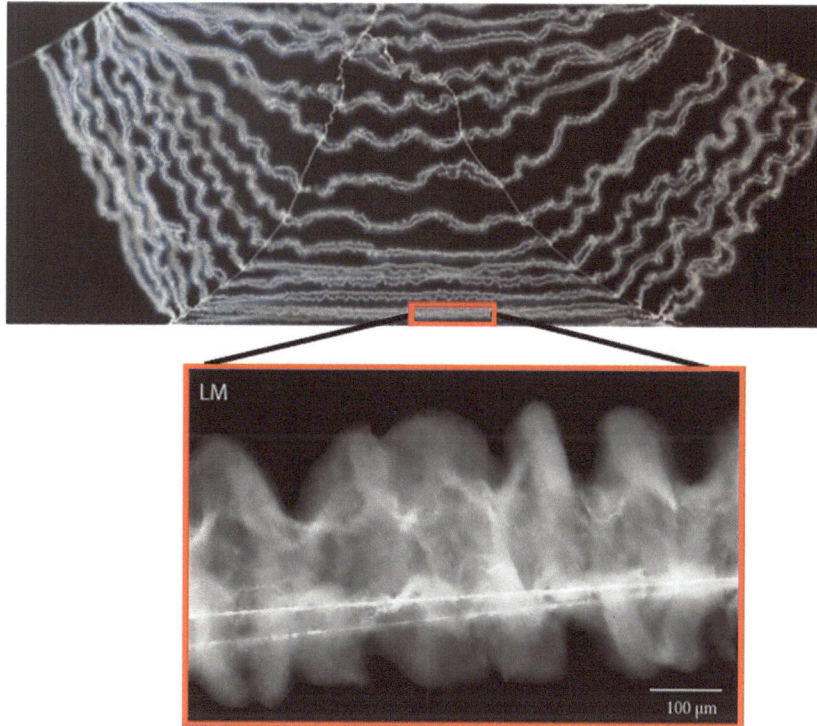

FIGURE 29. CRIBELLATE SILK SHOWING EXTREMELY FINE FIBRES SURROUNDING A LARGER AXIAL FIBRE.

Analyses of the cribellate silks in spiders from ancient families such as the Deinopidae (species from which include the net casting spider, *Deinopis subrufa*), the Desidae (which includes the house spider *Badumna longinqua*) and Austrochilidae (including the cave spiders *Hickmania troglodytes* and *Thaida pecularis*) suggest that the morphologies of the threads are always similar, at least within the bulk of the thread across species. The exception being threads in the spider *Kulkulcania hibernalis* (from the family Filistatidae), which are distinctly rounder and less matted and, presumably as a consequence of these morphological differences, less sticky than the threads made by all of the aforementioned cribellate spiders[121].

The net casting spiders are an interesting group of spiders as they have a unique way of using their cribellate silk. Rather than building an orb web that hangs from vegetation or another structure at night, the spider constructs a net resembling an orb web that is reduced in size which they hold open under stress using their forelegs. When its insect prey moves within the vicinity of the spider, it uses its legs to extend the net outward and orientated toward the insect before draping the net over the victim. The stickiness of the cribellate fibres entraps and immobilizes the insect enabling the spider to attack and consume it. This process is not entirely a passive sit and wait process. The cribellate silk reflects UV and blue light (Figure 31), which renders the net visible and attractive to certain types of nocturnal prey[123]. Like other cribellate silk-producing spiders, these spiders use the calamistrum on their forelegs to 'brush' the cribellate fibres on extrusion from the cribellum, which induces a charge onto the spun fibres, having the effect of fluffing the silk up and significantly enhancing their stickiness to an insect's cuticular surface.

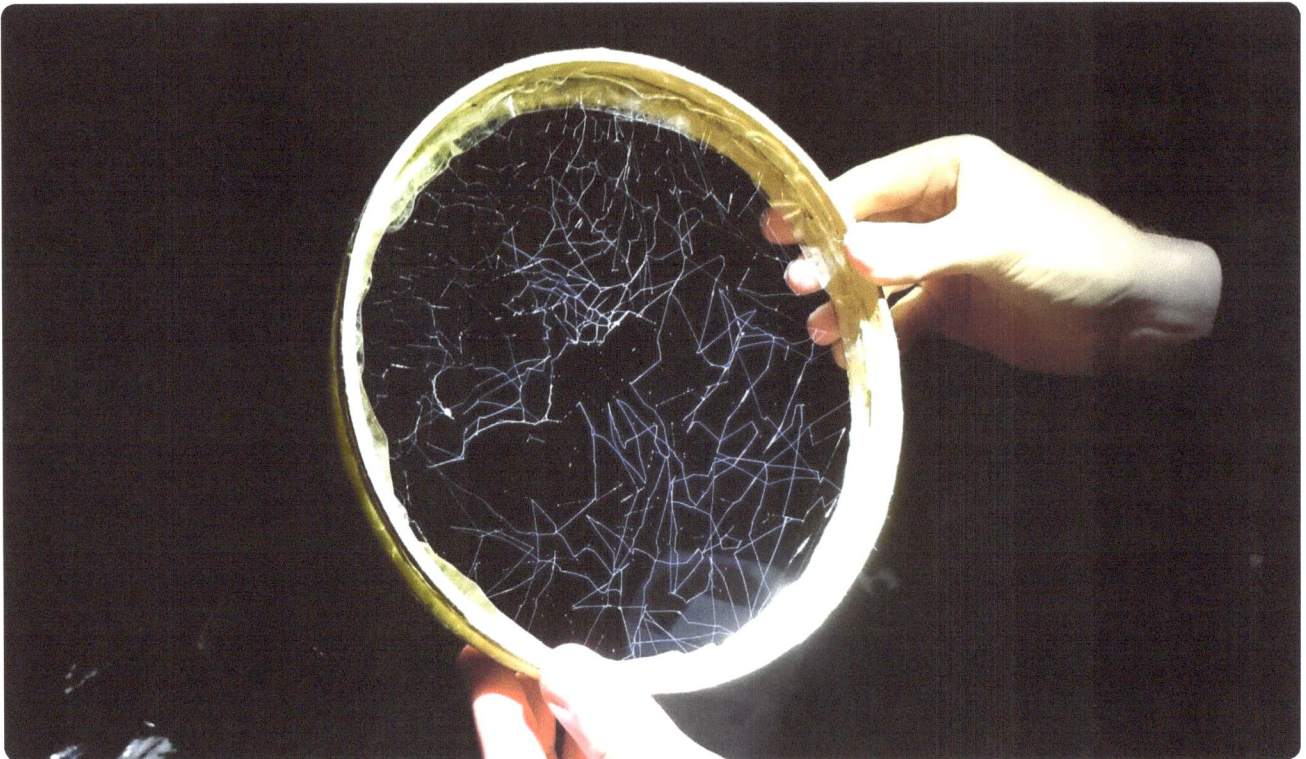

FIGURE 30. CRIBELLATE THREADS OF THE TASMANIAN CAVE SPIDER *HICKMANIA TROGLODYTES*, COLLECTED ON A BAMBOO RING, SHOWING ITS CRIBELATE SILK THREADS TO STRONGLY REFLECT UV/BLUE LIGHT WHEN ILLUMINATED UNDER TORCH LIGHT.

A cribellate silk spidroin has been isolated from mature females of a cribellate spider, *Tengella perfuga*, called CrSp. The amino acid sequence of this spidroin differs from all other spidroins, especially in the repetitive region, with amino acids that are traditionally found in abundance, such as glycine and serine, not being particularly high in abundance (although alanine is still relatively abundant). On the other hand, amino acids such as valine and leucine are in abundance in CrSp. Furthermore, there are no dominating repeating motifs in CrSp. This is consistent with the prediction that the silks do not conform into crystalline β–sheets, rather they form α- and/or so-called R- helical protein secondary structures[124].

The cribellate threads have a relatively high stiffness but low extensibility. Orb webs containing cribellate threads are thus much less capable of absorbing high impact prey. Whether or not wetting the cribellate threads induces changes in the web's mechanical performance remains contentious, with conflicting evidence found in different studies that have used different cribellate spiders. An equally contentious hypothesis posits that wetted *Kulkulcania hibernalis* cribellate threads retain droplets of water due to Plateau-Rayleigh instability acting on the thread surface, which accumulates where the silks aggregate into large or small knots[125]. The authors of these studies suggest that this water holding phenomena by cribellate silks enhances thread adhesion and might somehow be regulated by the spider's spinning.

Exposure to high humidity or mist can improve the adhesiveness of cribellate silks. However, rather than water accumulating at the surface of the thread, alternative studies show that water can infiltrate the proteins of cribellate silk and the underlying axial fibre and affect its secondary structures and thus its adhesive performance. This mechanism seems consistent with the idea that viscous silk did not suddenly evolutionarily appear but came about through the sequential modifications of cribellate threads with a gradual degeneration of the cribellum.[126]. Whether any hydroscopic salts are also associated with cribellate sticky threads to facilitate atmospheric water uptake is not known, but studies thus far have not found evidence for it.

Why the cribellum was lost

If cribellate silks perform just as well as the sticky viscid spiral silks, as they appear to do, and are adaptable across environments in a similar way to the viscid silks, then why were cribellate silks abandoned for viscid sticky silks in so many lineages of spider? An exceptionally high energetic costs of cribellate silk production has been posited as one explanation for the evolutionary replacement of cribellate silks with viscid spiral silks, with the viscous silk thought to be produced much faster, while being better at remaining intact across a wider range of conditions.

This simplistic explanation for the broad evolutionary shift towards viscous silks is however a fraught idea on many grounds. Firstly, there is evidence of flagelliform-like silks appearing in the silks of a range of cribellate spiders. This finding means that the switch over was likely to have been quite early on in spider evolution, perhaps before aggregate glues ever became available to spiders.[127] Secondly, many of the compounds found dissolved within aggregate glues seem unlikely to be acquired via the spider's diet and so need to be synthesized afresh metabolically, adding a previously ignored cost to the production budget of aggregate glue. On the other hand, we know that many orb web spiders consume their webs at the end of each day which, presumably, promotes the re-acquisition of most of the aforesaid compounds in their pure form. Thirdly, aggregate glues, owing to their high dependency on water for adhesion, often lose their functionality within a day out in the open, while cribellate threads can remain sticky for several days regardless of the conditions they find themselves in. Finally, recent spider evolutionary trees that have been built based on cutting-edge genome and transcriptome sequencing techniques show that orb webs containing viscous spirals have been repeatedly abandoned over evolutionary time[127]. This seems to suggest that viscous sticky spirals are just as, or perhaps even more, costly to produce and maintain as cribellate silks. Accordingly, the evolutionary shift in most spiders from using cribellate silk as the capture silk in webs to using viscous spiral threads as the capture silk, along with the subsequent degeneration of the cribellum as a spinning apparatus, remain unknown and await deeper investigations.

Other spider silks

Ribbon silk: flattened flexibility

Spider ribbon silks, particularly those produced by ancient recluse spiders such as *Loxosceles arizonica* and *Loxosceles laeta*, exhibit unique properties that set them apart from other types of spider silk. These silks are composed of very long, thin ribbons, with widths ranging from 2 to 4 micrometres and thicknesses of no more than 40 nanometres. Unlike the cylindrical threads spun by most spiders, the flat ribbon structure of recluse spider silk offers intriguing opportunities for detailed study and practical applications.

The morphology of the ribbon silk has been investigated using advanced microscopic techniques[128], which have revealed that it is highly uniform and possess remarkable electrostatic properties, which likely aid in prey capture. The silk's molecular structure consists predominantly of glycine and alanine in nearly equal amounts, forming motifs that adopts a β-sheet structure. This composition and arrangement confer the silk with exceptional strength and toughness.

Ecologically, spiders of the genus *Loxosceles* utilize their elastic silk ribbons to construct retreats and capture prey, employing a unique spinning mechanism. Unlike orb-weaving spiders that produce cylindrical threads, *Loxosceles* spiders extrude their silk through a slit-like spinneret, creating flat ribbons. This process is facilitated by a specialized gland that is much like the major ampullate gland of orb-weavers, which appears to be yet another fascinating instance of convergent evolution in silk production across different spider species.

The adhesive properties of the silk ribbons are particularly noteworthy. The combination of their extreme thinness, stiffness, and ability to conform to surfaces endows them with unmatched adhesive capabilities. Additionally, the presence of tiny, dot-like 'bumps' on the surface of the ribbons further enhance their stickiness[128].

Ancient spider silks: Threads though time

The extremely ancient mygalomorph spider *Linothele megathoides* builds large sheets as retreats. Recent research* has recently found this spider's silk to exhibit a moderate amount of stiffness and strength, but relatively low

---

* In a paper that is not yet published between myself and collaborators at the University of New South Wales, Deakin University, Australia, Arizona State University, USA, and the Universidad de la República, Uruguay.

extensibility and toughness compared to other spider silks. Its Young's modulus, or stiffness, is comparable to spider dragline and silkworm cocoon silk, but its lower tensile strength and extensibility suggest it does not perform as well under high stress, so it seems like it is not utilized within high-stress environments. Interestingly, almost half of the spider's silk is composed of the amino acids alanine, glutamic acid, and serine, with very little glycine, which is unusual for a spider silks. The toughness of *Linotheles megathoides* silk is reasonable given it lacks glycine motifs, which is a feature usually associated with tough spider silks. The primary structures that enable this silk to achieve its properties is somewhat of a mystery, and some of my colleagues are planning to examine more mygalomorph spider's silks to work this out.

The Long History of Silk Use

Silk industries of today

There is an enormous range of applications of silk in modern industries, from fashion to biomedicine, from cosmetics to recreation. Its natural sheen, smoothness, and breathability endows silk with extreme comfort, a pleasing feel, and visual appeal. The clothing and apparel industries exploit these properties to create fabrics for high-end fashion, alluring lingerie, and a range of luxurious accessories. Its ability to regulate temperature provides comfort in both warm and cool climates. It is, accordingly, a great choice for versatile wearables. Additionally, its hypoallergenic properties cater to individuals with sensitive skin, further enhancing its desirability in clothing such as pyjamas. These properties, along with its general popularity, mean many silk garments are relatively expensive compared to those made from other materials.

In biomedicine, silk's biocompatibility and biodegradability have opened avenues for its use as sutures, in tissue engineering, and drug delivery systems. Silk fibroins can be engineered into various forms such as films, hydrogels, and scaffolds, which support cell growth and tissue regeneration. This adaptability makes it invaluable for developing advanced and highly adaptive medical devices and treatments. Furthermore, silk's ability to become functionalized by the addition of bioactive molecules allows for targeted drug delivery, enhancing the efficacy of silk-based therapeutic interventions.

Artists and Designers also heavily use silk today, owing to its unique properties, particularly its ability to absorb dyes and its delicate yet durable nature. Silk is thus a favoured medium in most textile art, painting, and sculpture, where its lustre and texture can be harnessed to create more intricate and vibrant works.

Beyond these applications, silk has nevertheless started to become used in fields like electronics, robotics, and optics, where its strength, flexibility, and biocompatibility offer the potential for innovative uses such as within flexible and portable electronic devices and bio-active sensors. I will get into all of these examples in time. For now, here is a brief summary of silks use over historical time.

Sericulture: From emperors to eco-enthusiasts

Sericulture is the production of silk from farming and harvesting large numbers of silkworm moths. It is an extremely large industry, being practiced in at least 30 countries around the world, with China, India, Japan, South Korea and Brazil being the primary silk producing countries. It employs about 8 million people around the world and produces over 70 tonnes of fabric annually.

It is rather surprising to learn, as I did when researching for my Uruguayan lectures, just how far back in history sericulture dates. Indeed, exactly when it started is not well known. Most historians of the subject would estimate it to date back around 5000 years, with some archaeological evidence turning in dates of over 8000 years or more. Nevertheless, many of the most ancient examples are probably sporadic harvesting by the Egyptians and Chinese for burials and paper making. Essentially there is no true date for the origin of Sericulture. What is clear, however, is that it likely originated in China at least as far back as 2700 BCE.

One legend has it that silk harvesting was inadvertently discovered around that time by one Lady Leizu, the wife of the Chinese Yellow Emperor, Huangdi. It is said that she was sitting under a Mulberry tree drinking tea when a silkworm cocoon fell into her boiling beverage and the silk unravelled. Upon removing around a kilometre of soft, fine, shiny thread from her cup, she realized that these cocoons could be harvested to make ultra-lush fabrics, and thus domestication of the Mulberry silkworm moth began and Chinese Sericulture was soon established. Whether this story is true or not, the supposed timing of it coincides well with historical accounts of a remarkable increase in the use of silk in ancient China, albeit with that use being strictly confined to Chinese royalty and aristocracy. For over a thousand years, silk cloth was used exclusively by wealthy Chinese to pay taxes, fines, wages, and buy political favours. There is also some archaeological evidence of silks being used heavily in lavish burials, and other ceremonies. Silk also became increasingly popular with Buddhist monks for the creation of robes and embroidery, and it remains so among them today.

For around 2000 years the Chinese managed to keep Sericulture to themselves. How they managed to do so is not known. Nor is it known how their secret ever got out. One legend posits that around 200 BCE a Persian monk visited China and witnessed the silkworm moth larvae creating their impressive cocoons and saw how the cocoons were spun into yarn, whereupon he cunningly stuffed a hundred or so of the cocoons into the hollow of a walking cane. He returned to Persia via India and the Middle East revealing Chinese secrets along the way. While there is no evidence of this legend's truth, the supposed timing again coincides well with the rapid spread of Sericulture outside of China into India, the Middle East, and the Roman Empire, as well as to Japan and Korea. There is some evidence that India had Sericulture for almost as long as the Chinese, but they used a different assortment of silkmoths and harvesting processes. These included the Tasar (*Antheraea mylitta*), Muga (*Antheraea assamensis*), and Eri (*Samia cynthia*) silkworm moths, which still form a part of India's silk and Sericulture industry today. Nonetheless, these moths were never domesticated to the same extreme extent that *Bombyx mori* is, which does not exist at all today outside of silk farms.

By the first century AD silk was immensely sought after across Eurasia as an extremely luxurious commodity. As a consequence, multiple terrestrial and maritime trade routes between the Far East and the major Persian cities were established. These trade routes would soon spread over the preceding centuries into Europe and Africa. Eventually they became an immense arterial network of east-west trading channels supporting countless hubs and cities with new outposts consistently appearing along the various causeways that would one day become collectively known as the "Silk Road". While food, spices, animals, gems, other forms of clothing, dyes, and antiquities, were among the many items traded along the entirety of the Silk Road, it was silk fibres, and silken twines, fabrics, and the clothing made from them, as well as the moths that produce it, that were the primary goods of interest all the way along it.

It cannot be understated how much of an impact the Silk Road had on shaping the world it traversed. These impacts included the establishment of modern-day cities, and the spread of foods, goods, art, writing, scholarship, and religious and other ideas, from the east to west and vice-versa. One of the biggest impacts of the Silk Road, nevertheless, was an invisible one: disease!

Caravans moving along the Silk Road with goods intended for far flung markets undoubtedly carried many inadvertently hoarded passengers, which likely included rats and other rodents. These passengers carried fleas, which carried the bacterium *Yersia pestis*, the bacterium responsible for the Bubonic Plague, and other deadly diseases. Whether the Plague carrying bacteria originated in or near Europe and spread east or originated in the Far East and spread west is not known, but repeated outbreaks along the Silk Road trade routes were known to occur. The most infamous and terrifying outbreaks of Bubonic Plague occurred in the 14th century. This one, of course, became known in Europe as the Black Death. Records suggest that the bacterial strain responsible for the Black Death originated in China in 1338 and spread across the Silk Road steadily over about nine years, reaching Europe in 1347. It then proceeded to kill about half of the continental population in two short years. The tale of the spread of the Black Death demonstrates how profoundly silk and its remarkable lustre and other properties has impacted world events. On the flip side, many world events have themselves had significant impacts on Sericulture and the use of the silk products it has created.

If you take a second to think about what you might consider the most impactful inventions to emanate from of the Industrial Revolution, the steam engine and subsequent locomotive would probably come in at number one. Then next, according to many historians, would be the somewhat less celebrated Jacquard loom. While mechanical looms were in operation at the time of Joseph-Marie Jacquard's invention of around 1801, the Jacquard loom differed from other mechanical looms because the looming instructions for specific types of fabrics were stored on punch cards that were fed into the machine. The significance of the Jacquard loom on modern society thus includes the invention of the punch card system of storing information, which ultimately led, albeit after some false starts, to the first computers. Being operated by a single card feeder, the loom significantly saved on labour costs facilitating an immensely enhanced output per unit cost by manufacturers. The cards could be fed by an unskilled labourer meaning there was no longer a need for highly skilled weavers within the garment producing factories. This generated considerable anger among the weavers who, inspired by a figure from several decades earlier by the name of General Ned Ludd*, took to breaking into factories and smashing up the machines that supposedly put the skilled workers out of employment.

The other important impact of the Jacquard loom on society was a revolutionization of the European silk industry. Prior to the loom's appearance, silk products were prohibitively expensive in Europe owing to the sheer number of trading hands the silkworm moths and/or their silk products would have crossed as it travelled east along the Silk Road. The greater demand of silk products among nineteenth century Europeans as a consequence of new wealth generated by the industrial revolution was accelerated by the invention of the loom and other technologies that streamlined the processes of cultivating, spinning, weaving, and finishing, silken fabrics. While these events created a short-term boom in Sericulture in Europe, the industry, nevertheless, soon experienced another invisible threat.

Louis Pasteur was undoubtedly one of the great scientific figures of the nineteenth century. The benefits the world has reaped from the practices of immunization, microbial fermentation, and pasteurization, as a result of his work has been widely recognized over the years. However, his impact on the European silk industry is much less celebrated.

FIGURE 31. EXAMPLE OF A DISCOLOURED AND SPOTTED MULBERRY SILKWORM WITH PEBRINE DISEASE.

By 1845 silkworms throughout Europe were suffering from a severe epidemic. Diseased larvae were not feeding and, as such, were experiencing stunted growth, meaning they were not maturing toward adulthood and were thus failing to build cocoons. Furthermore, most of the larvae were looking thin and a sickly cream or brown colour, and developed distinct black spots on their off-white bodies (Figure 31).

The problem was at first attributed to a fungus contamination of the Mulberry leaves that the silkworm larvae were feeding on. Nevertheless, when measures at eradicating the fungus from leaves failed to cure the silkworms, the French Ministry of Agriculture appointed Louis Pasteur to investigate and curb the epidemic. Pasteur determined that the larvae were infected by a kind of parasitic fungi called microsporidium, and the disease, that has since become known now as Pebrine Disease, was transmissible. The eventual eradication of Pebrine Disease by the implementation of strict sanitation procedures in 1865 saved the European Sericulture industry. However, other events, such as the opening of the Suez Canal, meant that a greater quality and quantity of silk could be easily imported from the East, and the size of the European industry severely declined soon afterwards.

There were some efforts made in nineteenth century Europe to develop a spider silk production industry. In one instance, a magistrate from Montpellier, France, called Francois Xavier Bon de Saint Hilaire, began collecting

---

* Ludd supposedly went around smashing up factories as a form of protest in England in the 1780s. The movement accordingly became known as the Luddite movement.

the local orb web spiders and silked them using a machine of his own invention. Some garments of a sort were made, most notably a pair of stockings, but a spider silk industry never eventuated. It was clearly too costly and too difficult to collect and draw silk from enough spiders for the venture to last. Enterprising Europeans and Americans rather turned their attention to developing synthetics, or "artificial silks".

Artificial silk: when science began imitating nature

Attempts to create a synthetic silk date back to around 1855, as far as is known, when a Swiss chemist called George Audemars attempted to spin a cellulose-based fibre by cultivating and digesting Mulberry tree bark. The spinning method was later refined by Sir Joseph Swan, one of the many nineteenth century inventors attributed with discovering the incandescent light bulb. Indeed, it was his desire to find a suitable material to act as the light bulb filament that drew Swan to work on an artificial silk. What Swan did to produce his threads was force a liquified cellulose mixture through extremely fine nozzles to produce an extremely robust fibre. His method, without him knowing, resembled the natural silk spinning methods of silkworms and spiders. Much more than Audemars' method did anyway. Swan displayed his fibres at an International Inventors Exhibition in London in 1885 and drew many compliments. By the 1890s the isolation of nitro-cellulose by Hilaire de Channot inspired the French to open manufacturing plants to mass develop an 'artificial silk'. A rayon-like fibre was subsequently produced but it was highly flammable so could not be developed into a useful fabric. The production process was modified in Britain by Charles Frederick Cross which subsequently led to the discovery, in 1894, of a fibre known today as viscose or viscose rayon.

By the 1920s viscose rayon was flooding the market, and viscose-derived rayon fibres could be purchased for around half the price of traditional silk fibres. Silk still had its niche in the marketplace, however. Its lustrousness and durability made it more popular than the "artificial" fibres among the wealthy upper classes in Europe and the United States. That changed, nevertheless, with the development of nylon, the so-called "miracle fibre", in 1931 by Wallace Carothers at DuPont in Delaware, USA. While viscose rayon and silk were both natural products, derived from tree bark and silkworm cocoons respectively, nylon was completely synthetic. Its production was made possible by manipulating certain types of polymer derivatives, taken from the petrochemical industry, which came to be called thermoplastics.

A polymer is essentially a repeated chain of chemical units called monomers, the amino acid units making up a protein is an example that I have indeed already mentioned herein. The rapid manufacture and manipulability of nylon made it useful for almost any application. Moreover, its smooth appearance was comparable to that of silk. The first nylon product on the market, around 1935, was a nylon-bristled toothbrush. It later became high fashion as the material for women's stockings. During World War II nylon became heavily manufactured for use in parachutes and parachute cords. Now swimwear, electronic parts and machinery, fishing line, tennis rackets, carpets, and much more, are made from nylon or very similar petrochemical derived synthetic polymers. It truly did live up to its label as a "miracle fibre" throughout the 20th century. Nowadays much cheaper, but poorer quality, nylon is used than that initially developed by Carothers.

The sheer enormity of the depth and breadth of the modern-day manufacture and use of nylon fibres and other petrochemical based thermoplastics, such as polyester, has a dark and burdensome downside, in the form of pollution and environmental damage as well as severe energy and resource inefficiency. The creation of polyester fibre, for instance, uses seventy million barrels of oil each year, and the manufacturing processes creates many types of pollutants that seep into waterways and bioaccumulates in aquatic organisms. Added to that, extremely high temperatures are required in the production of polyester to induce it to polymerize, which of course burns significant amounts of polluting fossil fuels. To finish the material and form fibres, the polymers are treated with a multitude of additional noxious solvents that themselves often end up in the environment during or upon manufacture.

The environmental damage done in the everyday use of these materials is even more striking. For instance, a 5-kilogram wash of clothing derived from polyester fabrics releases over six million microfibres which remain in the environment and bioaccumulate or float around in waterways to attract further organic pollutants and bacterial blooms. Additionally, the mass production of plastic materials forms textile wastes that contributes to around 5% of the world's landfill problems. Overstocked clothing fabrics get either burnt or contribute to even more landfill. Once into landfill or water drainage systems, the fibres seep into soil and ocean ecosystems where they persist for tens to thousands of years and become incorporated into ecological and human food webs. As a result, research that focuses on improving the longevity performance of non-polluting natural fibres, such as silk, has undergone a recent revitalisation.

A Natural Solution: Looking to Silk to Inspire Sustainable Solutions

Switching from plastic to silk products presents itself as one compelling solution to the ongoing global pollution crisis. Unlike synthetic plastics, which persist in the environment for centuries, silk is a natural material that breaks down rather harmlessly in the environment. If all plastic fibres can be substituted with silk-based alternatives the release of plastic pollutants into the environment can be minimized, or even eliminated, and the volume of microplastics entering ecosystems around the globe can be drastically reduced. Silk's versatility of use, biodegradability (which means it can be broken down once in the environment by bacteria), and general safety, make it an excellent candidate for replacing a wide range of plastic products, if a cheap manufacturing method could be devised.

Silk can be selectively processed to create coatings that attract or repel water, depending on the application, which is ideal for producing biodegradable microcapsules. These microcapsules can be used in pharmaceuticals, cosmetics, and agricultural chemicals, ensuring a controlled release of active ingredients while eliminating the environmental hazard of traditional plastic microcapsules. Silk's non-toxicity also means it can be safely used in food and medical applications, offering sustainable and health-conscious alternatives to plastics. Tools and equipment like whipper-snipper blades could be made from silk composites. The nylon blades that are currently being used break up and fragment into microplastics and these persist in soil and water systems for centuries. Silk blades would provide the necessary durability for effective trimming while being biodegradable once cut and disposed of.

Other materials could be manufactured from recycled silk or silk that was considered too low grade for use in textiles or other high-performance materials, thus broadening the potential range of silk products and making the use of silk more economically viable. Silk recycling and manufacturing infrastructure could be readily adopted globally, thus reducing dependence on specific regions for supplying the silk. The widespread adoption of silk-based alternatives, accordingly, would promote new economic activity based around silk, and not only help mitigate plastic pollution but also support the development of a circular economy where waste is minimized, and natural resources are efficiently used.

Harnessing silk using sericulture practices is not particularly cost effective compared to the development of synthetic polymers, however. Cutting edge genetic engineering technologies, such as recombinant and transgenic silk production, polymer engineering, biomimetic and bio-inspired engineering designs, and advanced fibre spinning techniques, nevertheless offer alternatives that could eventually enable the production of more environmentally safe fibres at competitive costs. This is particularly important for the large-scale production of spider silk-like materials as spider farming or natural fibre harvesting have long proven prohibitively time consuming and enormously expensive. Indeed, there are now many research groups and start-up companies around the world developing the kinds of technologies necessary for creating synthetic silks, and they are accordingly becoming progressively closer to producing ready-to-use high performance, robust, silk-like fibres and materials.

Fibres of the future

$\mathcal{I}$t is reasonable to wonder why, given its amazing properties, a spider silk based super high-performance rope, extremely strong packaging materials, high performance clothing, implantable medical prosthetics, space elevator, or the highly sort-after light weight bullet proof vests, or indeed a spider man suit, has not yet been developed. What have the spider silk researchers, designers and entrepreneurs been up to in all the years since Saint Hilaire? You might also rightly state that the environmental benefits of using this material instead of thermoplastics should make such developments an extremely high priority.

The reason why we don't have these long-promised materials and designs lies in the technological limitations currently being experienced and the prohibitive costs involved. But things are changing really fast.

The vast bulk of today's silk production still relies on sericulture. As in the past, modern sericulture comes at high labour and handling costs, limited potential for real growth, and a high prevalence of diseases among silkworm colonies. These issues prevent the silk industry from becoming more widespread in the immediate future and it means that silk fibres are unlikely to become available at drastically reduced costs any time soon. Nevertheless, an increase in the accessibility of new genetic and polymer engineering technologies will significantly reduce the costs and has excellent potential for enabling the scale-up of silk and silk-like fibre production for an ever-expanding range of uses.

Genetic engineering is the modification of the genetic material, or sections of deoxyribose nucleic acids (DNA), of living organisms in order to manipulate in some way what it expresses genetically. To this extent silkworms may be considered, because they in fact are, a genetically engineered organism. The engineering has just occurred over a very long timespan (5000+ years). You might call it ancient genetic engineering. Modern genetic engineering however involves directly altering the gene sequence of a stretch of DNA in an organism and inserting it into another organism, or merging it with another gene sequence, a so-called recombinant sequence, that had been cut or sequenced from another organism using a special enzyme. In most instances, genetic engineers manipulate the DNA of some kind of bacteria or yeast to circumnavigate the ethical dilemmas surrounding altering of the genetic structures of animals, and because it is much easier to use viruses to transfer DNA into yeast and bacterial genomes.

Attempts to produce genetically engineered silkworm, bee, and spider silks have so far mostly involved inserting recombinant gene sequences into a bacterial (most commonly the gut bacteria *Escherichia coli*, with other common bacteria including *Bacillis subtilis* and *Salmonella typhimurium*[128]), yeast (usually *Pichia pastori*)[129], plant (for instance tobacco) or animals like moths, hamsters, mice, or goats, as hosts. All with variable degrees of success. The premise is that the host will express the inserted silk genes and produce silk proteins that can then be spun into silk-like fibres. Because the mechanical qualities of spider silk, such as its strength, extensibility, stiffness, and toughness, are considered greater, thus superior, to silkworm silk, and because of the difficulties farming and collecting silks from spiders, most silk genetic engineering efforts have so far focused on engineering types of spider silks. I will do that too from here on unless stated otherwise.

Bacteria and yeast microbial silk factories

The very first recombinant spider silk protein was constructed in 1995[129], when parts of the newly described MaSp1 and MaSp2 proteins were created using the *E. coli* bacterium as the host. Since then, there have been many reports describing new strategies for producing recombinant silks for research and commercial applications. For spider silk, the research has focused so far on recreating the two major ampullate proteins, MaSp1 and MaSp2, in the lab since a mix of these proteins seems to be essential for the generation of the high strength and high elasticity found in major ampullate silk.

As I have stated earlier, the major ampullate proteins have N- and C-termini that help the silk assemble and stay soluble, while their repetitive domain in the middle of the protein is key to the silk's flexibility and toughness. However, making these proteins in other organisms, like bacteria, is challenging due to their size and complexity, but new techniques are improving the process.

The larger size and considerable amount of repeated alanine units in the MaSp1 repetitive domain render it harder for the various types of hosts to express and secrete than MaSp2, and laboratories have coped with this and other problems in a variety of ways (which I will get to later on). One problem with making recombinant proteins is that they generally cannot attain the same high solubilities in water as the naturally expressed spidroins do. This becomes a significant problem when trying to spin the proteins into fibres, as I will eventually explain.

The protocols used for creating the spider silk proteins by recombinant technologies are essentially the same as those for generating other proteins, such as human insulin and somatropin, and involves, firstly, extracting a small circular piece of DNA, called a plasmid, from a bacteria or yeast cell. A section of it is cut and replaced within a section of the silk protein gene. The modified plasmid is then re-inserted into the bacterial or yeast cell, which reproduces the plasmid containing the spider silk protein gene, expressing the gene as it exponentially multiplies over a few weeks[130].

Several laboratories have utilized recombinant technologies to replicate spider silk genes and create proteins. One has produced a close to full replication of the spidroin MaSp2 using an *E. coli* host.[131] Nevertheless, most efforts to produce a recombinant form of MaSp1 have so far only produced subsections of the protein of rather low molecular masses. One particular group from China[132] has managed to produce a near full length MaSp1 protein and spin it into a robust fibre. This was achieved by making the bacteria grow within a glycine enriched broth so that they are forced to over-produce metabolic glycine. Such results have nevertheless not yet been replicated elsewhere.

Experimental evidence suggests that it is essential that highly water-soluble recombinant proteins of close to full length need to be produced in order to spin long mechanically viable threads[67]. However, so far, the recombinant technologies have not been able to create MaSp1 proteins within bacterial hosts, at least not any with molecular masses close to those of full-length proteins.

Researchers are now diverting their attention toward finding a way to 'stick' together subunits of recombinantly expressed MaSp1 proteins. One example of a MaSp1 subunit is the 4RepCT protein, a protein that comprises a four unit repeating sequence and a C-terminal domain[133]. This approach means it is not necessary that the bacterial or yeast host express a full-length protein[67]. An alternative approach to recombinant expression that is becoming widely explored is 'chimeric' or 'fusion' protein synthesis, wherein full length spidroins are expressed by joining segments of *MaSp1* and/or other spidroin genes together to express the protein segments, or where feasible, whole chimeric protein chains. This technique gives researchers the option to combine a multitude of complementary genes from different species into a single plasmid for expression.

Since it is the interactions between the N- and C-terminal domains of the spidroins and the ionic and pH environments within the silk gland, or spinning columns when considering synthetic proteins (which I'll discuss later), that induces secondary structural formations in the fibres during spinning, any incomplete or re-arranged sequences within the repetitive regions will be inconsequential on the structures and mechanical properties of the spun fibres. The detailed compilation of *MaSp1* gene libraries that have been assembled for many spider species in databases such as the European Bioinformatics Institute and NCIB Genbank, and the aforementioned 1000 spider species Silkomics database, means it has become possible to digitally create a mixture of different chimeric *MaSp1* gene sequences. Such digital sequences then can be used to order a plasmid from a supplier of Synthetic Biology products for subsequent expression in a chosen host. I therefore expect the ability to synthesize full length MaSp1 and MaSp2 proteins to be just around the corner.

There is some speculation that using yeasts as the recombinant host produces larger quantities of complete spidroins.[128] For example, the biotech company Bolt Threads has used yeast hosts to manufacture spider silk proteins *en mass* for weaving into a fabric called Microsilk™; a clustered type of polymer, termed a 'block copolymer', comprising of different spidroins, including near full length digitally-derived MaSp1 and MaSp2 proteins[134].

Alternative spider silk proteins to the major ampullate proteins MaSp1 and MaSp2 have been produced by recombinant technologies utilizing bacterial and yeast hosts. These include partially sequenced replicas of the MaSp2-like proteins ADF3 and ADF4*, Flag derivatives, derivatives of the proteins TuSp1 and PySp2,[135] and an array of cysteine rich proteins (CRPs) that are associated with major ampullate silk[136]. One laboratory at Dalhousie University in Canada, for example, has been successful at producing a 230 amino acid repeating unit of PySp1 and a 200 amino acid repeating unit of recombinant AcSp1 within bacterial hosts[137]. These proteins were developed upon determination of the full sequences for these proteins within the orb weaving spider *Argiope trifasciata*.

Plant hosts: can we farm fibres?

Spider silk proteins have been developed by combining spider silk genes with plant genes. In so doing, turning the plants metaphorically into factories for producing silk proteins. Plants that naturally accumulate high concentrations of proteins within its seeds, leaves, tubers or roots, such as potato, tobacco, soybean and *Arabidopsis* have been used to produce MaSp1 and MaSp2 with varying levels of success. An advantage of using plants over bacteria or yeast for recombinant silk gene expression is that higher concentrations per unit effort are attainable within plants. One experiment, for example, found that 2% of all soluble proteins expressed by a tobacco plant represented transgenically expressed MaSp1[138].

Recombinant gene duplication involves creating multiple copies of a specific gene within an organism's genome, while transgenic expression refers to the introduction and expression of a foreign gene from another species into a genome to produce new traits or characteristics. Depending on the proteins expressed, a combination of recombinant gene duplication and transgenic expression might be employed to scale up the yields. A combination of methods has been utilized to maximise *MaSp1* expression within tobacco and soybean for example[128]. In one particular study, a Flag protein from the spider *Trichonephila clavipes* was produced in transgenic cauliflower by promoting the expression of additional self-splicing molecules called inteins[139].

Utilizing plants for the recombinant production of spider silk proteins is promising because it leverages the natural protein-producing capabilities of plants, making the production process more sustainable and potentially more cost-effective compared to traditional methods involving bacteria or yeast. In addition, it opens up new possibilities for large-scale production. Plants can be grown in vast quantities, and their cultivation can be scaled up relatively easily. All of this could make it feasible to soon produce spider silk proteins on a commercial scale.

---

* ADF3 and ADF4 are just so named because they were sequenced from the spider *Araneus diadematus*.

Animal hosts: Going beyond moths and spiders

Inserting spider silk protein genes into animals such as silkworm moths, hamsters, mice, or goats, for expression within specialized cells, such as in the silk glands of insects or mammary glands in mammals, is complex, involving multiple cell transplantations and rearing individuals across generations (see Figure 32 for example). The process of transferral of the genes is also complex as the genes have to be transferred into the genome of the animal using a viral vector. This requires recognition of specific components of the target genome. As an example, a vector called *Bombyx mori* Nuclear Polyhedrosis Virus (BmNPV) has been utilized to transfer recombinantly amplified chimeric spidroin genes into the genome of Mulberry silkworms for expression and secretion in their silk and salivary glands.

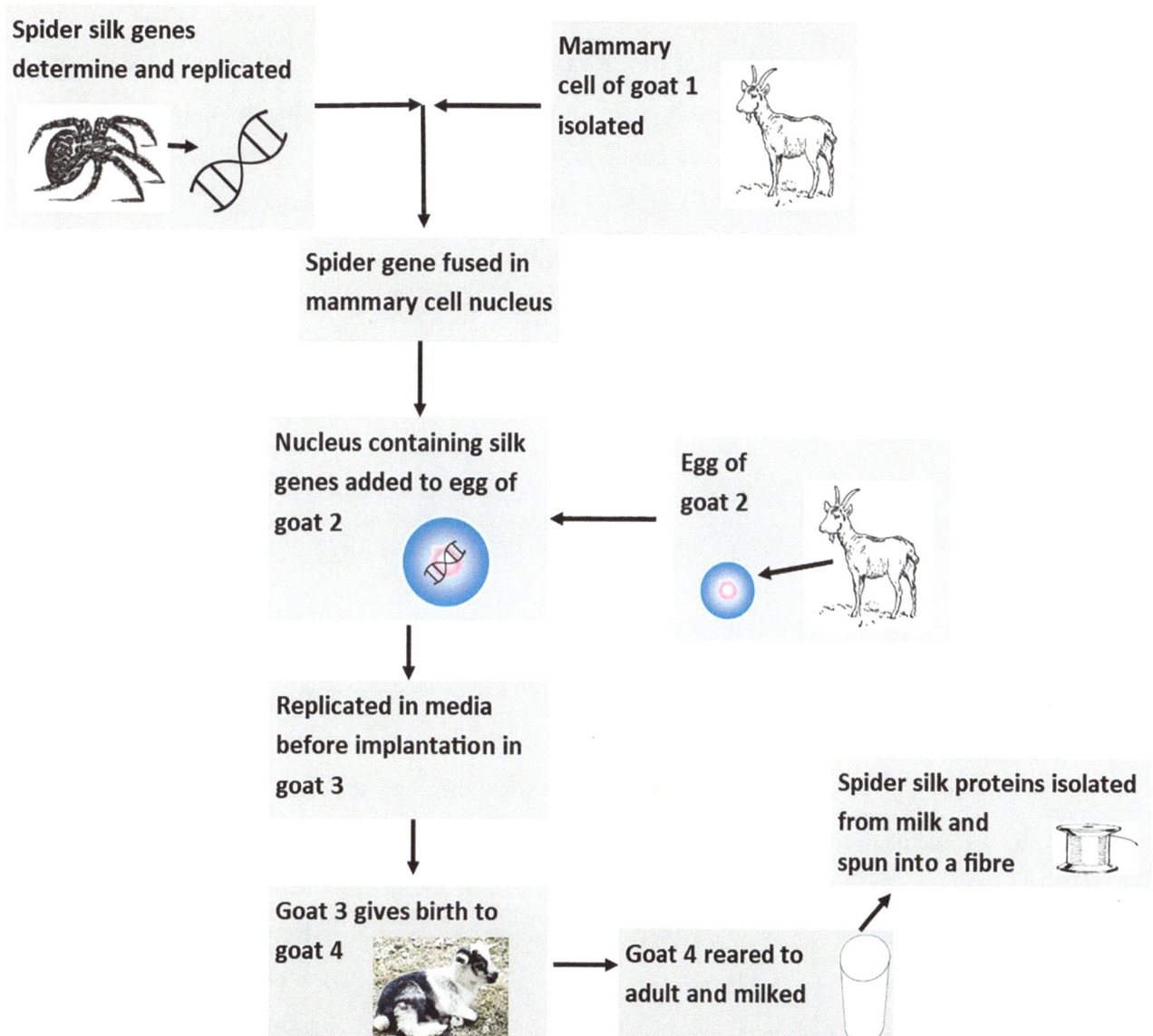

FIGURE 32. METHOD OF INSERTION AND EXPRESSION OF GENES TO MAKE SILK PROTEINS WITHIN ANIMAL (GOATS IN THE EXAMPLE) CELLS.

Several laboratories have produced a form of spider major ampullate silk within the glands of transgenic silkworm moths. Kraig Biocraft Laboratories, for instance, uses genetically modified silkworms to produce different versions of spider silk-like fibres, known as Dragon Silk™ and Monster Silk™. These silks integrate spider silk genes into the silkworm genome, thereby combining the silk-producing efficiency of silkworms with the superior mechanical properties of spider silk.

Most of the efforts so far have produced blended fibres of silkworm and spider silk. The mechanical performance of the blended fibres are comparable to native spider silk but the spidroin yields in the moth silk are relatively low[140]. One method that increases the spidroin yield within transgenic silkworms is the insertion of a promotor protein which controls the enzyme-driven cutting and pasting of spidroin genes at specific locations. This allows for the insertion of specific transgenic silk genes at specific locations within the silkworm genome.

An alternative method is to use cutting edge technologies, like CRISPR-Cas9, to edit the silkworm's genes to enable them to express spider silk proteins. The isolated "clustered regularly interspaced short palindromic repeats" (CRISPR) sequences, along with the "CRISPR-associated protein 9" (Cas9), allows for extremely precise gene editing to efficiently incorporate spider silk genes into the silkworm genome. In one study, silkworm heavy and light chains were edited to express small sections of MaSp1 and MiSp1, respectively, that were then assembled to build fibres of strength and extensibility approaching that of native spider major ampullate silk[141].

These studies demonstrate that there is enormous potential for high yield spidroin expression and the production of high-quality fibres using transgenic *Bombyx mori*. However, all of the aforementioned problems associated with commercial sericulture (i.e. labour and handling costs, slow growth potential, the prevalence of diseases) persist, so the ability to scale up production at low cost is still going to be limited.

Over the past 10 years or so, numerous experiments using various transgenic mammals, including mice, hamsters, sheep, and goats, have been undertaken with the aim of inducing the expression of recombinantly amplified spider silk proteins at extremely high yields within the animal's milk. Experimenting with mammals, nevertheless, yields results much more slowly than similar transgenic procedures with silkworms as they require the passing of the silk genes from mother to daughter before subsequent expression and secretion within the milk[142]. Accordingly, even though this research is now well over 10 years old, it is still not known whether or not mammals are a route to mass production of spidroins or spider silk fibres.

Of the transgenic mammal programs that have operated over the years, the most successful and well-known is the spider silk transgenic goat program at Utah State University in the USA. Two lines of transgenic goats have been bred at Utah State, one producing milk containing MaSp1 and another producing milk containing MaSp2. The spidroins produced are not of a completed sequence and not large enough to be spun into fibres resembling spider silks in their properties[143]. So far, the proteins seem to be primarily used for experimentation but they hold some potential for being made commercially available. Spider silk fibre production thus could scale up via this route should it ever become possible to induce the goats to express full sequence spidroins.

I expect that of the animals used for transgenic production of spider silk, silkworms appear to hold the most promise for producing fibres with properties that mimic those of spider silks. Indeed, it has already been used commercially by some biotech companies, such as Kraig Biocraft.

Using silkworms to mass produce spider silk makes sense because silkworms already spin massive amounts of fibrous silk and they are an evolutionary closer relative to spiders than any mammalian host, meaning their cellular machinery is likely to be somewhat similar to that of a spider. Thus, it is expected that silkworms should have the metabolic capability to be able to read the long spidroin gene sequences and fully translate them into functional fibres. Indeed, there has been much frustration in getting mammalian or plant or bacterial hosts to read and produce full length, or even close to full length, spidroins. Furthermore, silkworms are quite easily, at least compared to spiders, reared and farmed *en masse*.

An advantage of goats or other mammals as hosts over other animals is that they can potentially produce massive amounts of the target proteins in their milk. However, rearing and housing them in large numbers is complex and time consuming. Furthermore, the molecular weights of the proteins produced have so far been too low for making high performance fibres. Nonetheless, now that a transgenic colony is established, perhaps they can be utilized as a source of proteins for additional modifications using polymer engineering or other future advanced technologies.

What is certain is that companies wishing to create high performance functional materials for purposes such as biomedicine, electronics, space applications, military and other types of wearable applications, or spiderman suits, need to be experimenting with native spider silk to get their readiness to such a level that they are able to manufacture a product as soon as large masses of high-quality recombinant spider silk fibres become available.

Scaling it up: production from the petri dish to world stage

Despite sounding like a very 21st century concept, polymer engineering is a rather old technology. Since synthetic rubber was first made in the 1800s, most of the materials we have used over the past two centuries have been manufactured using polymer engineering processes. Indeed, viscose rayon, nylon, polyester, and other thermoplastic derivatives, have all been created using types of polymer engineering techniques for many decades. Given this, you might expect that if there were a way to mass produce silks or materials with silk-like physical properties then surely it would have been done by now using such technologies. Nevertheless, for much of the past two centuries the technologies available to properly manufacture silk-like fibres has been out of the reach of polymer engineers. Recent advances in the field, however, might be about to change this.

As I had described already, recombinant and transgenic protein production technologies are enhancing our capacity to attain different kinds of proteins in the laboratory. There is, however, no point attaining high yields of any kind of silk proteins by these or other means if your ability to stitch the proteins together into a silk fibre is substandard or non-existent. This is pretty much the primary impediment still facing all current attempts at creating spider silk-like fibres at an industrial scale today. Nonetheless, advances in a multitude of polymer creation technologies, including electronic gene sequencing, chimeric protein sequencing, protein assembly, and fibre dry, wet, and electro-spinning technologies, and three-dimensional printing technologies, all show enormous potential for enabling the mass production of spider silk-like materials to soon be realized.

The broadest and most ambitious scientific project stemming from the 1990s was the Human Genome project. While a thorough understanding of the human genome was undoubtedly invaluable for clinical and other social reasons, probably the next biggest achievement of the project was the development of the newfound capacity to map whole genomes, or subsets of genomes, of organisms at an over 1000-fold reduction in the costs. Accordingly, a wide range of animal genomes have since been sequenced. As might be expected, the genomes of the most economically valuable plants and animals were some of the first to have their genomes fully sequenced. Among these, of course, was the silkworm moth *Bombyx mori*[144]. Following this the silk-specific genomes of the bagworm moth[49], and a few iconic spiders, including the golden orb weaver, *Trichonephila clavipes*[54] and Darwin's Bark Spider, *Caerostris darwinii*[145], have been sequenced.

Whilst genome sequencing and mapping have meant that details about silk genes and the proteins they code for are becoming more available to researchers, harnessing the information to build proteins in the laboratory via recombinant or transgenic routes remains difficult; for reasons that I have touched on.

I have already stated how the digital sequencing of genes is increasingly being used to rapidly construct recombinant proteins. There are however a few other genetic technologies that have recently been developed or are being developed that are real game changers.

Nanopore sequencing is one such technology that is changing the fields of genetic analyses and gene technologies at an exponential rate. Using this technology the rapid sequencing of entire genomes can now be achieved. This is done by DNA or RNA strands getting passed through a flow cell containing an immense number of nanopores —tiny protein-based or synthetic holes—all aligned in an orderly arrangement called an array within a membrane containing a slight charge. Instead of allowing only single nucleotides to pass, entire strands of nucleic acids move through the nanopores. As each of the nucleic acid's nucleotides pass through it causes characteristic disruptions in the membrane's current, which are detected by the sequencer. Specialized algorithms within the sequencer interpret these current changes in real time to determine the sequence of the DNA or RNA.

Next-generation sequencing (NGS) is another technology enabling the rapid sequencing of genomes. This technology works by rapidly and cost-effectively sequencing millions of DNA fragments in parallel, providing high-resolution insights into the structure and function of genomes. When it comes to synthetic spider silk production, NGS enables scientists to precisely sequence and analyze the silk-producing genes of spiders, which can then be synthesized or engineered in other organisms, such as bacteria or yeast. This technological advancement has real potential to revolutionize synthetic silk creation by facilitating the production of silk with properties, such as increased strength or elasticity, that are tailored for specific applications.

There is palpable excitement surrounding these technologies. The market value of companies such as Illumina Inc. that invest in them is expected to triple from around 9 billion USD to around 27 billion USD by 2032. Researchers and biotech companies are equally excited by the potential applications of nanopore sequencing and NGS, from personalized medicine to pharmaceutical to environmental monitoring. Its ability to provide rapid, accurate sequencing data is impacting many fields, including genomics, molecular biology, and synthetic biology. The promise of real-time sequencing data opens new possibilities to produce synthetic fibres, not to mention the possibility of tracking evolutionary changes in silk properties across the invertebrate species that secrete them. The pace at which this technology is advancing suggests we might just be on the brink of a new era in biological research and biological innovation.

Even with such technologies, ascertaining full length DNA sequences that code for extremely long silk proteins, such as MaSp1, is still difficult, but accurate segments of proteins can now be sequenced and stuck together within a bacterial plasmid using proteins appropriately called "sticky ends". Using such techniques, relatively small components of MaSp1 and MaSp2 have been produced electronically (and have been thus called eMaSp1 and eMaSp2) and engineered into bacterial genomes. The resultant proteins that these bacteria have expressed have been subsequently stitched together and spun into fibres with secondary structures and mechanical properties that are remarkably similar to naturally spun spider silk, although in just rather small amounts thus far[146].

Chimeric proteins are being prepared in laboratories by fusing together stable fractions of different protein genes within virus vectors for transferral into bacterial or yeast plasmids. The bacterial or yeast host subsequently expresses and assembles a full-length fused protein molecule. This method has become popular for attaining spinnable lengths of MaSp1 proteins from spliced sequences of *MaSp1*, 4RepCT, and other parts of similar spidroins. What is more, silks that have been spun from chimeric proteins perform very well in tensile tests, with some attaining strength and modulus values nearing natural spider silks[147].

An alternative method that has been advocated recently is attaching a series of peptides, the molecular building blocks of proteins, that consist of multiple amino acids along a scaffolding polymer, such as polyurethane, to form what is called a pseudoprotein or facile protein. In one experiment, pseudoproteins were engineered and used to dry spin a silk-like fibre that achieved a reported toughness comparable to that of *Argiope trifasciata* aciniform silk when tensile tested[148].

Advances in spinning: Getting the process right

Running solubilized polymers, such as silk proteins, cellulose, or thermoplastic polymer melts, through a spinning column and forcing them out of some kind of nozzle or artificial spinneret at high pressure as the means of producing new fibres has advanced significantly since Joseph Swan first made a synthetic cellulose fibre over 150 years ago. Advances in the engineering of the spinning columns and output nozzles have enabled manufacturers to tightly control factors such as the amount and quality of the polymers dissolved in the spinning dope, the viscosity or thickness of the dope during spinning, and the forces acting on the dope as it flows through the spinning column. By using a spool to further pull on the fibres as they are extruded from the nozzle, manufacturers have been able to introduce a straining flow force that applies friction in a similar way as that applied at the spigot when silk is spun by a spider, moth, or other arthropod. Post-spin chemical or physical treatments of the fibres such as immersion in ethanol or methanol, or other dehydrating fluids, during wet spinning, or exposure to some kind of drying column during dry spinning, might be applied to draw out water and fix the proteins in position and prevent the fibre from falling apart. Primarily, for a range of reasons, wet spinning is currently more commonly used to spin silk fibres than dry spinning.

Attempts to produce silk fibres using modern wet and dry spinning methodologies date back to at least the 1960s. Most of the experiments were performed by dissolving and re-spinning silkworm silks. Experiments spinning reconstituted silkworm silk fibroins (or RSF) involves degumming cocoons to remove the sericin layer and spooling the remaining fibres. The fibres are thereupon dissolved in a chaotropic* salt solution such as a concentrated lithium bromide solution. Once dissolved, the silk proteins are concentrated further by dialysis or by precipitating them out and re-dissolving them.

The process of producing reconstituted spider silk proteins from silk fibres is much different. There is not any sericin layer, so degumming is not needed. As an alternative to chaotropic salts, which do not dissolve the silk, spider silk gets dissolved in volatile organic solvents such as hexafluoro-propanol (or HFIP) or N-methylmorpholine- N-oxide (NMMO). These agents are expensive and volatile, and this is one of the reasons why spinning reconstituted spider silk fibres has been done infrequently.

The mass production of fibres from recombinant proteins with a mechanical performance that matches those of naturally spun silks is not quite achievable but, as discussed already, we are getting rather close, albeit only producing small amounts of fibre. The ultimate prize, or the so-called "holy grail of chemical engineering" however, is to find a technological platform for spinning large quantities of high performing recombinant or transgenic fibres that perform as well as spider major ampullate or aciniform silks.

Unravelling the Recipe for a Perfect Fibre

Once a concentrated solution of reconstituted or recombinant silk proteins is prepared as a dope, it can be pumped through a syringe into a fixing solution under high pressure to form a fibre. A constant high pressure on

---

* A chaotropic salt is one that can, when at sufficiently high solution, induce substances like as biological macromolecules like silks to dissolve in water.

the syringe can be achieved by using a syringe pump. Modern syringe pumps can be programmed to enable the researcher to tightly control the pressure applied to the syringe and so regulate the rate of flow of the dope into the fixing solution.

The concentration of the proteins within the dope needs to be within specific limits (8-16% being considered the optimal) to flow with enough viscosity for the tapering down of the column width to place shear forces on the solution in the same way that the silk dope experiences shear force in the spinning duct of a spider or silkworm to subsequently form a fibre. Recently, researchers have been running the dope through complex spinning columns etched into a non-toxic silicon-based polymer called polydimethylsiloxane (PDMS), or three-dimensionally printed silicon-based lab-on-a-chip devices. These work by increasing the shear forces on the dope, either directly or via the introduction of a focusing fluid that flows over and around the dope, or by altering the geometry of the column in some way to induce additional shear forces. The focusing fluid induces the flow rate to increase and a thinning out of the dope before it is forced out of the column. Experiments have shown that the angle and direction by which the focusing fluid meets the dope is guided by the way the spinning column or lab-on-a-chip devices are engineered. The geometry of the column affects the amount of shear applied on the dope and, subsequently, the dope's viscosity within the column, induces draw down taper whereupon the proteins self-align, and the silk dehydrates and solidifies into a fibre.

The focusing fluid can be any solution that has a carefully controlled pH and ionic strength and is designed to closely mimic the natural physical and chemical conditions within a spider's silk gland. By precisely manipulating these conditions, the focusing fluid induces alignment and folding of silk proteins at the correct stage of spinning, thereby facilitating the transition from a soluble dope to an insoluble fibre. The fluid's composition is tailored to create a shear flow environment that helps in stretching and orienting the protein molecules, thereby producing fibres with enhanced mechanical properties. Using focusing fluids thus allows for the controlled and reproducible production of synthetic spider silk.

During wet spinning the fibres are drawn into a coagulating bath (which may run from a few centimetres to up to 1 metre long) of concentrated (around 95%) methanol, ethanol or isopropanol, or a concentrated salt solution such as aqueous ammonium sulfate. This solution dehydrates and solidifies the fibre and locks the proteins in position which enables the formation of the protein secondary structures. Upon coagulation, the fibre is drawn out of the bath before being either immersed again within a second coagulating bath and then spooled, or immediately spooled in air. Experiments with recombinant spider silks often add water to the coagulating bath to slow down the rate of coagulation to induce further alignment in the proteins and to increase the extensibility of the fibre [149]. Spooling the fibres at a rate of around four times faster than the syringe pump's flow is required to add enough strain on the fibre to ensure that the proteins fix in position and the fibre does not crumble[150].

To dry spin silkworm or spider silk from their dopes additional treatments are needed. This usually involves mixing positively charged particles, such as calcium ions, into the dope solution to generate electrostatic conditions conducive to drawing out water from the fibres upon spinning. Dry spinning is probably a mimic of the natural silk production process. It is also a cleaner method of silk fibre formation as it does not involve fixing fibres in noxious dehydrating solvents such as concentrated alcohols. However, since the dehydration stage is rapid during dry spinning, it is much more difficult to produce sufficiently long and stable fibres[151].

Factors affecting the dynamics and efficiency of dry and wet spinning include the amount and concentration of the dope solution, the pressure applied by the pump, the channel length and width, the material or materials from which the chip and/or spinning column were manufactured, the composition of the coagulating bath, and the post spin reeling speed. Each of which have been shown to affect the properties of the fibres that are ultimately spun.

For successful wet or dry spinning, a silk dope first needs to be prepared of sufficient protein concentration. Rheological analyses (which just means the analysis of liquid flow) suggest that the protein concentration ideally should lie between 8 and 16%. Furthermore, the protein molecular weights need to be at least around 280kDa, which is close to the full length of most fibroins and spidroins. Making a dope of too low a concentration does not produce fibres, while dopes with too high of a concentration solidify prematurely inside the spinning column and so the dope either flows extremely slowly or the spinning column clogs up. Dopes of just under 8% protein, or with proteins with low molecular weights, will probably form fibres but they will be unstable or fragile.

Silkworm silk fibres that have been wet spun from dopes with low concentrations of RSF undergo significant necking when placed under any kind of strain. Adding a concentrated calcium salt solution, such as a calcium chloride solution or some other stabilizing agent, to the dope enables the spinning of quite stable fibres when using a dope with a protein concentration of slightly lower than 8%[152].

When necessary, additional shear forces can be induced on the dope as it flows through the spinning column by etching corners into the column at certain locations, or flowing a very highly viscous focusing fluid around it. Polyethylene glycol is the most commonly used focusing fluid for this purpose[152]. The focusing fluid can prevent aggregation of the proteins within the column, enabling high concentrations of high molecular weight dopes to be utilized thereby enhancing the robustness of the spun fibres. A mildly acidic focusing fluid can be created by adding dilute acetic acid to the polyethylene glycol solution and this may help to create the pH gradient needed to trigger protein aggregation and initiate the assembly of the protein structures[152]. For this reason, a coagulating bath of slightly higher acidity than the focusing fluid is commonly used for wet spinning silk of fibres; as it helps promote protein secondary structure formation and locks the protein chains in place.

Altering the pressure applied at the syringe pump, either by varying the rate of pumping or the volume of fluid being pumped, has the effect of altering the rate at which the dope flows through the entirety of the microfluidic system; from syringe to spinning column to nozzle. The effects of increasing the pressure applied at the pump can be difficult to quantify. One experiment[196] used a range of focusing fluids and multiple spinning columns to moderate the fluid flow rate through a microfluidic system and found that increasing the flow rate produced fibres of a decreased diameter.

The use of PDMS and/or three-dimensional printing lab-on-a-chip devices has given researchers the ability to produce spinning columns that closely resemble the geometries of silkworm or spider spinning ducts and spinnerets. This means that all the natural physical conditions experienced by silk dopes inside a gland can be replicated in the laboratory. It also means researchers can isolate parameters, such as the amount of tapering within the duct, shear-force induced fibre thinning, or draw down taper, to experiment on their relative influences on fibre properties.

Something that obviously differs substantially between PDMS, lab-on-a-chip, or other microfluidic devices used in fibre spinning and the spinning processes in real silkworm or spider glands is the materials that the respective surfaces are composed of. The various devices are produced from a silicon polymer, such as PDMS or glass, or similar, while real glands are made of living tissue. This inevitable incompatibility between real and artificial systems may explain why it is so difficult to create artificial fibres with properties matching those of their natural counterparts. Not only does the material used affect the spinning process but it is nearly impossible to create appropriate pH and ion gradients across a silicon-based spinning chip without the introduction of additional micro-channels to the column, which will inevitably lead to altered flow rates and create a range of additional uncontrollable influences over dope viscosity[198]. On top of this, it is not possible to mimic the ability of living systems to adjust dope viscosity on the fly or control the inward and outward flow of water across the glandular cells to control fibre dehydration rate.

Simulating the controlled dehydration has been a substantive difficulty that has been worked around in many ways. When wet spinning, the spun fibres are drawn through some kind of coagulating fluid. This fluid is usually a concentrated alcohol such as methanol, ethanol or isopropanol, a concentrated salt, or other dehydrating solution. The purpose of the coagulating bath is to dehydrate the fibres and lock the silk proteins in position as it solidifies into a fibre. The chemical nature and concentration of the coagulant and the amount of time the fibre is exposed to it also affects the properties of the fibres that form.

Using concentrated alcohols as the coagulant dehydrates the fibre faster than using salts does. As a consequence, the β-sheets crystallize more rapidly in fibres coagulated in alcohols and tend to align more randomly. This results in stiffer and less extensible silks than might have been formed had the fibre been dehydrated slowly using salts. Of the salts, aqueous ammonium sulfate seems to be the most popular. One experiment using transgenic goat spidroins showed that using ammonium sulfate as the coagulating fluid improved extensibility by around 100% and toughness by 10% compared to when using isopropanol as the coagulating fluid[143]. The drawback of using ammonium sulfate is that it is extremely volatile and noxious if inhaled by the operator. As an alternative, researchers have used tetrahydrofuran (THF), a more benign organic solvent than any of the alcohol or ammonium sulfate solutions, and found it enhances the fibre's extensibility and toughness by around 50% compared to methanol or ethanol.[152]. Water may also be added to the coagulating bath to dilute the solutes and to slow down the rate of coagulation or to induce the reuptake of water and reduce spider silk back to its supercontracted "ground state"[149].

The amount of time the fibre is exposed to the coagulant can be modified by running the fibres through shorter or longer baths, depending on how readily the fibres dehydrate and form structures. Sometimes two or more coagulating baths are used sequentially. They may be of increasing concentration to enable the fibres to become drier on each sequential submersion to ensure the structures fix in place. On the other hand, the fibres might be exposed to the most concentrated solution first, so the fibres can form and dry out before the properties are manipulated by the sequential exposure to increasing quantities of water[153]. In some instances, a series of coagulation baths have been used to expose the fibres to solutions of increasing acidity in an attempt to replicate the pH gradient of a spider or silkworm silk gland, as it is the gradient, rather than pH itself, that triggers the formation of the protein secondary structures.

In addition to exposing the fibres to different chemicals for different amounts of time, other post-spinning modifications that might further influence the properties of wet spun silks include exposure to heat and/or some kind of post-spin stretching. A combination of heat and stretching has been shown to induce changes in protein secondary structure in reconstituted silkworm silk fibres[154]. How effective a specific treatment is at improving the fibre's performance depends partly on aspects of the dope* and the coagulant the fibres are exposed to. For instance, one experiment[155] found that, all else being equal, stretching wet spun spider silk fibres exposed to isopropanol stretched to 4-6 times their initial length and produced better performing fibres than those treated with methanol or ethanol, or those stretched less than 4 or more than 6 times their initial length.

Electro-spinning is an alternative means of spinning recombinant and reconstituted silks. It involves using an electrical charge to draw the polymer solution or aqua melt through extremely fine capillary tubules under high pressure. Consequently, extremely fine, sub-micrometre fibres are formed and sprayed out of the spinneret. Upon exposure to air the electro-spun fibres coalesce into an airborne aerosol. A collecting surface or plate intercepts the aerosol particles. Accordingly, electro-spinning is used to produce nonwoven materials, such as coatings, films, gels, sponges, foams, and powders. Electro-spinning may also be used to apply a coating of silk film, gel, sponge, foam, powder, or other material onto a surface during manufacturing processes. The density and properties of the silk coating will depend on the size, shape and surface temperature, concentration of the proteins in the dope feedstock, the pump speed (which also affects the viscosity of the feedstock in the spinning column), and the voltage and temperature under which the dope is drawn[156]. The main use of electro-spun silk materials today is for the development of transplantable medical devices, films, and gel-based scaffolds for applications such as cell culturing.

---

* Whether it consisted of particular fibroins/spidroins, their concentrations, molecular weights, amino acid compositions, and the number of monomer sub-units they were formed from.

Challenges remaining: Nature's secretes that are proving a tough nut to crack

There are still many challenges associated with using wet and dry spinning techniques to create synthetic silk-like fibres. For instance, a challenge for spinning spider major ampullate silk-like fibres lies in developing technologies that incorporates MaSp1 and MaSp2 proteins in a way that they are combined and distributed throughout the fibre to produce the desirable ratio of β–sheets, β–coils, β-turns, 3_{10}-helices and other secondary structures throughout the fibre. The development of new digital engineering technologies, such as e-textile technologies, additive manufacturing, and three-dimensional printing applications, could potentially alleviate these kinds of issues but it remains to be seen.

The technologies might enable protein structures to be programmed and inbuilt into printable materials. Experiments have already been done to recreate whole spider webs with the same material properties across their surfaces as real spider webs[157], and digitally created silkworm silk fibres have now been three-dimensionally printed for incorporation into wearable fabrics[158].

As I explained earlier, silk proteins are large, with intricate sequences and precise folding patterns that are an essential component of their unique mechanical properties. While there has been some success in producing silk proteins in host organisms like bacteria, yeast, or plants, the process of spinning these proteins into fibres that mimic the natural spinning process of spiders or silkworms remains an elusive goal. Current attempts have repeatedly produced incomplete or incorrectly folded proteins, which lack the desired strength and elasticity of native silks.

Another challenge is the economic viability of producing recombinant silk fibres on a large scale. The current methods for producing silk proteins through recombinant technologies are costly and yield relatively small quantities of silk, making large-scale production impractical at the moment. Moreover, the purification process required to obtain high-quality silk proteins adds to the overall production cost. These economic barriers hinder the development and commercialization of silk-based products, despite their promising applications in various fields, such as biomedical devices, textiles, and high-performance materials. The development of cost-effective and scalable production methods is going to be crucial for realizing the full potential of recombinant silk technologies.

Should it become possible soon to spin silk-like fibres in the laboratory at a low cost, several challenges remain in the development and commercialization of usable products. One primary concern is ensuring the consistency and quality of the fibres, as variations in production can lead to differences in performance characteristics. Additionally, integrating the fibres into existing manufacturing processes for textiles or medical devices may require significant adjustments, incurring further costs and development time. Regulatory hurdles must also be addressed, particularly for biomedical applications, where safety and efficacy standards are stringent. Finally, market acceptance and consumer perception play crucial roles; educating consumers about the benefits and potential applications of recombinant silk products will be essential for successful market penetration. Overcoming these challenges will undoubtedly be vital to fully harness the benefits of recombinant silk fibres in various industries.

Current and Future Applications for Superfibres

There have been claims in different types of media about the wonders of some forthcoming applications for silk products*, and breakthroughs in developing artificial materials with properties that resemble natural silks. Some are real, some will be plausible soon, some are still a long way off, and some plain fanciful. I will navigate through some of the claims to give a sense of what has and has not been done, and where the field is heading. I will, where applicable, describe how harnessing specific properties has permitted certain applications or is preventing a proposed application coming into fruition. I will review some of the more ideal scenarios that might eventuate should we accomplish the ultimate prize (or holy grail) of finding a means for spinning large quantities of silk-like products.

Silks in Medicine

Silk has a long history of being used in medicine. This association primarily comes about because the high tensile strength, anti-bacterial, anti-inflammatory, biocompatibility biodegradability, haemostatic, and non-cytotoxic (cell killing), properties of silk products lend themselves very well to medical uses. The Greeks and Romans used to bundle together spider and other silks to treat the wounds of soldiers. A sixteenth century French surgeon called Ambroise Paré first used silk threads to tie together severed blood vessels. As there are too many historical examples to compile herein, the interested reader should see a recent detailed review of silk's many medical applications that appeared in the journal *Advanced Healthcare Materials* in 2018[159], or one of the many other fantastic reviews of silk's uses across the medical disciplines.

---

* Among these are claims of a spider silk-shrimp cuticle hybrid to replace plastics, spider silks being used as robot muscles, genetically modified silkworms creating spider silk-like threads, and the development of new high performance armour from genetically engineered silk fibres.

Given the rich history of silk in medicine, it is befitting that it is the medical field that has benefited the most from the modern-day resurgence in silk research and applications. Among the applications, products, and treatments that have recently become available are biodegradable sutures and wound dressings, a range of implants for replacing or regenerating different tissue types, prosthetics, and vehicles for the safe delivery of drugs and enzymes. So promising is the use of silk in the medical industry that at least 18 different companies are now producing silk-based medical products.* Indeed, I am constantly hearing about more companies moving into this field, and I would not be surprised if there were in fact more than 50 companies by the time this book comes out. Most products are made from reconstituted silkworm fibroins and sericin, although there are a few products derived directly from fibre sericulture or the reconstituted silks of spiders or insects other than moths.

There are only a few current applications of spider silks in the medical field. However, I believe that this will become a new frontier and those companies investing now in research using natural spider silks will get an advantage over those not yet investing in such research, as their applications will be prototyped using the gold standard native material.

Healing, Sealing and More

Dressing wounds to promote wound recovery is one of the earliest known medical applications of silk that dates at least back to ancient Greece. There is however anecdotal evidence of Australian and New Guinean Indigenous people using spider webs and silk to treat wounds for many millennia. It seems that silk's anti-bacterial properties and its permeability to water and air while covering a wound to prevent further contact and contamination make it an excellent dressing for many types of surface wounds.

A new range of silk-based wound coverings are now being experimentally developed. They are usually made from reconstituted silkworm silk mats, sheets or films structurally enhanced with polyethylene oxide, silver salts, or azide[†] ions.

Suturing wounds using silk threads dates back to around 150 AD. In the 1860s Joseph Lister, on finding the link between infections by microorganism and the spread of disease, started sterilizing silk sutures using carbonic acid which led to an increase in the survival rates from surgery. Today high performing permanent sutures are made from petrochemical derived synthetic polymers such as nylon or polyester. This is because these materials retain their strength over time so can allow wounds to gradually heal over days or weeks.

---

[*] According to Table 5 of Holland et al. 2018. There are without doubt additional companies that were not included therein and companies that have appeared since this publication.

[†] Azide compounds contain a negatively charged radical of three nitrogen atoms.

Most of today's biodegradable sutures are made from natural fibres such as silk, catgut*, or cotton. These sutures break down over time so do not need to be physically removed by a surgeon. They are preferred for repairing internal incisions or ruptures as non-degradable alternatives require further surgery to remove. Probably the most common product being developed by the 18 or more companies involved in developing silk-based medical supplies is silk-based sutures. The fibres can be relatively easily produced from sericultured silks or reconstituted fibroins. The flexibility of silk to change in property in different environments means that silk sutures can better cater for specific surgical needs, for instance wet suturing or small-scale surgical procedures, than other types of sutures. Their naturally low immunogenicity make silks attractive as a material for advanced wound healing applications or applications where the risk of adverse reactions to the sutures and other materials used are unacceptably high.

It is currently being explored whether electrospun silk nanofibres have potential as wound healing scaffolds. These nanofibres physically mimic the extracellular matrix and they provide a supportive environment for cell attachment and proliferation. Such scaffolds can be further functionalized with growth factors, antibiotics, or anti-inflammatory agents to enhance healing.

Additionally, silk hydrogels, derived from both silkworm and spider silk, are being developed as injectable wound fillers. These hydrogels can conform to irregular wound shapes and provide a moist environment, which is crucial for optimal wound healing.

This increasingly new range of silk-based wound coverings and other surgical applications is now being experimentally developed and prototyped. More will no doubt be done in this area in the near future. The majority of applications thus far are utilizing structurally enhanced reconstituted silkworm silk mats, sheets, or films. *Ex vivo* trials suggest that the sutures, dressings, and fillers are all effective at promoting rapid wound healing, just as long as they are applied in a dry and sterile environment, such as a hospital[160].

Implants, Tissue Culture, and Prostheses

There are many challenges associated with developing silk-based medical implants. These include minimizing the body's innate immune response against the implants themselves while retaining the structural and functional integrity of the transplanted tissues. Accordingly, any silk-derived biomedical material must be both biocompatible and able to bear similar mechanical loads as the original underlying structural collagen. A wide range of sericin and fibroin-based biomaterials have been successfully transplanted into model animals. These include skin, nerve, bone, tendon, muscle, ligament, eardrum and eye lens replacements, with most verified to retain their functionality well without inducing inflammation or any other antagonistic response.

---

* More accurately this should be called sheep or horse gut.

One proposed application is to use recombinant or reconstituted silkworm or spider silk films or membranes as a replacement skin. While there has been no successful development yet of a silk-derived human skin transplant, there have been some inroads made toward such a development. For instance, a full-thickness fibroin-based skin-like film has been electro-spun and grafted onto rat models. The "artificial skin" nevertheless lasted for just 18 days before it showed signs of deterioration[161].

Grafting silk skins onto rats is one thing, replacing large sections of human skin is however a much greater challenge owing to the larger surface area, complex geometry, surface topography, unique physical characteristics, and functional variability across the body. A deeper investigation of the functional compatibility of spider silk to human skin is thus needed before a biocompatible silk-derived human skin can ever become fully developed.

Much research has been done with animal models on the transplantation of silk fibres as nerve replacements and conduits* and the grafting of fibroin or spidroin films onto nervous tissue or silk threads and sheets as replacements for faulty myelin sheaths; the insulating layer surrounding nerve cells, and other nerve components. *Ex vivo* studies have shown that silk fibres and films are able to replicate the conductive function of nervous tissue[162]. Likewise, natural and artificial silks have been shown to accept and retain myelin cells[162]. Recently, a recombinant spider silk-myelin cell composite material was used to promote the regeneration of defective peripheral nervous tissue *in vivo* in sheep[163]. These studies accordingly suggest that it is rather feasible that we could see silk-based biomaterials used for human nervous tissue repair or replacement in the near future.

Silk's structurally hierarchical high tensile strength and general biocompatibility means that silk-based biomaterials are ideal for regenerating bone cells for grafts or replacements. The biomaterials used are created by allowing calcium phosphate crystals to grow and assemble on a gel or film scaffolding made of fibroin-hydroxyapatite or fibroin-tricalcium phosphate [164]. Experiments suggest that the regenerated bone materials can be implanted without inflammation or rejection, with one experiment using mice showing fibroin-hydroxyapatite bone replacements to promote bone recovery within 12 weeks[165].

Sericin and fibroin derived films, sponges, and hydrogels, have high permeability for oxygen and water vapour and, accordingly, support cell adhesion and promote cell growth. These materials have thereupon been used as a culturing medium for a range of body tissues[166]. Anterior cruciate ligament (ACL) cells have been regenerated on silk sponges knitted around a porous silk mesh and successfully inserted into pigs[167]. Silkworm and spider silk gels, films, and sponges have also been successfully used as scaffolds to support regenerated or replace ligaments, blood vessels, and skeletal and cardiac muscle, and cartilage-forming chondrocytes for insertion into animal models[156]. Researchers at the Ear Science Institute, Australia, and the Institute for Frontier Materials at Deakin University, Australia, have been developing, as a part of a program called the "Tympanic Membrane Project",

---

* A nerve conduit being a synthetic nerve replacement to hold a severed nerve in place until such a time that the nerve can grow back. It is important that the conduit is easily removed or degrades, and no non-nervous tissue grows into the gap between the severed nerve before or after the application.

transplantable ear drums using fibroin films. The drums attain, owing to the compliance of the underlying silk fibroin, a combination of high strength and stiffness, capacity to transfer vibrations at specific resonances, and high biocompatibility within animal models[168].

Silk biomaterials are also increasingly being used as therapeutic devices and prostheses. Silk is useful for limb prostheses as it has a high load-bearing capacity that is likely to be equal or better than any of the replaced body parts[156]. Being biocompatible and biodegradable, silk has started being used as temporary vascular prosthetics that can enzymatically degrade as the original vascular tissue regenerates[169]. To date, all of the silk prosthetics in use are made from regenerated silkworm fibroins. I nevertheless expect spider silk prosthetic development to become feasible soon given the recent advancements in genetic engineering, protein engineering, and spinning and printing methodologies. Again, I expect that it will be the companies investing early in research using native spider silk that will gain the edge in this soon to rapidly expand field of biomedicine.

Delivering Drugs Where They're Needed

Silk-derived drug delivery systems in the form of capsules and adhesive nanoparticles are becoming increasingly used. Both silkworm and spider silks have been used for making drug delivery capsules and nanoparticles. However, for reasons associated with their respective costs, and the general expectation that silkworm biomaterials perform better and degrade more slowly than spider or other silks, silkworm silks are used most frequently.

Drug delivery and release capsules as three-dimensional porous silk-derived membranes and films are now being developed. These capsules are broken down by digestive enzymes or stomach acid and the drugs they carry are effectively delivered into the body across the wall of the digestive tract. In most cases the silk-based membranes are engineered to be degraded by digestive enzymes by a two-step process. The first step is adsorption of the enzyme onto the membrane surface. This step is facilitated by the presence of a surface-binding molecule called a domain. The second step uses the enzyme to break up specific chemical bonds, called ester bonds, across the membrane surface[169]. A capsule that is too strong or stiff may be hard to break open so the capsule's strength and stiffness may be moderated during development of the membrane or film by diluting the fibroin or spidroin solution in glycerol. This serves to curtail the formation of the crystalline β–sheets in the developing membrane or film[170].

Silk-derived nanoparticles are often used to deliver or assist drugs in the treatment of cancer. The nanoparticles and anticancer drug combination can be injected together into affected tissue to take effect locally. The nanoparticles are produced from fibroins by a process called nanoprecipitation, where nanoparticles are forced out of a highly concentrated solution of some kind of polymer through a series of chemical reactions. Specific nanoparticles are then selected from the resulting precipitate for their ability to penetrate specific types of cancer cells and their capacity to slowly release the attached anti-cancer drugs. Among the types of cancers that the procedure is most well developed for is breast cancer. This is because the silk-derived nanoparticles have, for reasons that are not clear, a very high affinity for binding with cancerous breast tissue cells[171].

Engineering the Future

In addition to the many types of medical products, aids, and devices, each of which utilizes silk's unique mechanical properties within fibres, casts, films, or gels, in their own way, silks are being utilized to, or are earmarked to, create a multitude of other useful products. Among these are silk-based interfacial brain-borne sensory microchips that utilize silk's electrical conductivity properties and biocompatibility for use *in vivo* to help regenerate lost hearing, vision, or motor control[172].

There are also plentiful bioengineering applications, which includes developing silk fibres, proteins, or particles specifically to bind and tack on to chemicals through microfluidic and optofluidic (a technology platform that combines microfluidics with optics) systems, or the development of biochemical sensors for a range of analytical and practical applications such as the fluorescent labelling of nucleotides during genetic sequencing[173].

A group, at Swinburne University in Melbourne, has been experimenting with and manipulating the thermal diffusivity*, and thermal and electrical conductivity, of silk fibroins for potential use as micro-optical sensors and semi-conductive devices for advanced lithographics (where an image or text is drawn on a flat surface, usually stone or metal, with an oily substance), among other applications. The team worked out that the thermal diffusivity and electrical resistance of silk fibroins can be chemically manipulated, thus opening up the possibility of developing a new range of silk-based temperature, light, and electrically conductive materials and applications[174]. They have also trialled these applications using freshly spun native spider silks.

In addition to these advancements, researchers are exploring the potential for silk engineered products in environmental engineering. One promising avenue is the development of a silk-based filtration systems for water purification and air filtration. Due to its strength, elasticity, and biocompatibility, spider silk could be engineered to capture and neutralize various contaminants, including microplastics and airborne pollutants, offering sustainable and biodegradable alternatives to synthetically derived filters.

Another exciting application is in the field of energy storage and generation. Research teams around the world are investigating the use of spider silk as a key component in bio-based batteries and supercapacitors, leveraging its high surface area and conductivity to improve the efficiency and sustainability of energy storage devices. Initial trials of developing these new technologies using freshly spun native spider silks are showing enormous promise. It therefore seems highly feasible that high performance silk-based electrical engineering materials will be developed in the not too distant future.

---

* Thermal diffusivity is a measure of a material's ability to transfer heat relative to its ability to store it.

Adhesives: Silk's Sticky Solutions

The loss of adhesion as temperature or humidity fluctuates is one of many problems associated with modern synthetic adhesives. There is accordingly a lot of interest in research aimed at understanding the adhesive mechanisms of natural glues, such as mussel byssus and glow worm and spider gluey silks. Applications that could be developed from this research includes *in situ* tissue adhesives, toxin free food packaging adhesives, wet and/or underwater adhesives, and adhesives for electronics and robotics that are reusable in any environment.

The unique adhesive properties of spider aggregate and pyriform silk and their ability to function within a water or organic media renders them excellent candidates for inspiring the production of novel biomedical adhesives. Sealants and reinforcements for wounds, or for the repair of fractured bone and other tissue, are among an array of potential applications.

Recent studies have revealed that the adhesiveness of orb web spider aggregate glue is the result of extension dependent behaviour of the glue droplets (see the section on spider aggregate glue). These droplets are composed of an aqueous coating comprising of an array of salts* surrounding microscopic nodules made of glycoproteins. When the flagelliform and aggregate silks become extended, as they do when capturing flying prey in an orb web, the aggregate glue becomes viscous. This activates the glycoproteins, making them highly adhesive against insect cuticular surfaces. If an attempt is made to pull the sticky thread from the surface it quickly becomes stickier. As the insect moves the glue droplets deform like rubber and move along with it[175]. This is why it is virtually impossible for insects stuck in the web to wriggle themselves free from the web. What is more, the chemistry of the spider glues gives them such enormous adaptability that they can adhere to many normally "non-stick" surfaces, such as the loose scaly wings of butterflies and moths[176].

The mechanisms driving aggregate glue's ability to reversibly vary in viscosity and adhesiveness are now being investigated to develop a range of high performing synthetic "smart" adhesives that are adaptable and reversibly adhesive across different surfaces.

For many physical reasons, adhesion is extremely difficult to achieve at the sub-micrometre scale. Glueing cells together or implanting devices onto cells is exceptionally difficult to achieve, hampering the efforts of microsurgeons and similar micro-practitioners. At such scales glues are rendered ineffective and materials are stuck together by exploiting short-range attractive forces, such as electrostatic forces and van der Waals forces.

The stickiness of gecko toes and spider cribellate silk are examples of biological materials that effectively utilize these forces. They involve the actions of extremely fine (i.e. nanometres thick) hairs, called setae in the case of gecko toes, or nanofibres in the case of cribellate silks. Understanding the adhesive functionality of such

---

* Which are highly variable in composition among the different orb spider species, with factors such as the spider's habitat and diet effecting their functionality.

exceptionally fine setae and nanofibres at the micro or nanoscales is valuable for providing the inspiration for creating adhesives that can function at such fine scales.

To this end, a European Commission funded research project called "Biocombs4Nanofibres" was done to understand how spider cribellate silks achieve their adhesion at the sub-micrometre scale. The multidisciplinary team from across Europe is now instigating the development of active nanofibre-inspired adhesives. In investigating how cribellate spiders produce and secrete their cribellate silk they looked at the specific role of the spider's calamistrum, particularly the geometry and patterning of its ripples, in enhancing the silk's adhesion. Projects such as this open up opportunities to create artificial silk-like materials with an adhesiveness that can be selectively manipulated by a kind of tiny comb with laser embedded ripples that mimics the spider's calamistrum[177]. This comb can be used to brush across nanofibres in such a way that their adhesiveness can be selectively enhanced.

Another promising area of development with spider silk-based adhesives is in the creation of environmentally friendly, self-healing, adhesive materials. Researchers are investigating how the unique protein structures within spider silk allow it to recover its shape and maintain its adhesive properties after deformation, and how these properties could be harnessed to create new self-repairing adhesives. These adhesives could be used in a range of applications, from extending the lifespan of consumer goods to improving the durability of medical implants. Initial studies have demonstrated that when silk-based adhesives are damaged, they realign their protein structures to restore their original adhesiveness, a functionality which has significant potential for many useful applications.

Spider silk is also being investigated as to whether it might be inspirational for the development of reversible adhesives in adjustable bonding materials. These materials would copy the ability of spider silk to detach and reattach without losing adhesive strength, providing an innovative solution for industries requiring temporary fixtures, such as in electronics or robotics. Preliminary research has shown that by mimicking the molecular composition of spider silk, particularly its ability to form and break chemical bonds under specific conditions, engineers might soon be able to create materials that offer strong, yet reversible, adhesion.

Strong, Soft and Sustainable Clothing

Clothing and fabrics, as well as linens, shrouds, paper, stitching and strings, have been produced from silkworm cocoon silk for thousands of years. Spider silks have a less illustrious history as a fabric but there have been some products made over the centuries. The most well-known item was the "Spider Silk Gown" (Figure 33) created in the early 2000s by collecting silk from 80 million Madagascan golden orb weaving spiders over 5 years. The gown's creators, Simon Peers and Nicholas Godley, forcibly silked the spiders using a series of devices similar to that created by Francois Xavier Bon de Saint Hilaire. They simultaneously silked 24 spiders to create the individual yarns. The gown now resides within the Victoria and Albert Museum in London as an exceptionally expensive display piece only. Earlier, around 1900, a spider silk tapestry was developed in France. It was estimated that around a million spiders had to have been sacrificed to create the garment. The original tapestry had been lost and a recreation built by silking another several thousand spiders later on. The recreation now resides in the American Natural History Museum in New York.

FIGURE 33. A GOWN MADE FROM THE SILK OF 80 MILLION MADAGASCAN ORB WEB SPIDERS BEING DISPLAYED.

Recombinant and transgenic spider, bee, and other, silks are increasingly becoming incorporated into clothing in novel ways. Companies such as Bolt Threads, Kraig Biocraft, Polartec, AM Silk, and Spiber are developing lines of recombinant and/or transgenic fibres and have had success in transferring them into clothes, including a beanie, tie (made by Bolt Threads), jacket (Spiber), and shoe (AM Silk). Spider silks are also attractive as an inspiration for developing a range of high performance materials and textiles called "smart materials" or "smart textiles"*, or in some circles, "intelligent materials" or "intelligent textiles"†, where the reversible mechanical properties of silks are incorporated into the material so that it can 'sense' and react to environmental stimuli. My colleague, Patricia Flanagan, at the University of New South Wales, for instance, has created a kind of smart material inspired by the supercontracting properties of spider major ampullate silk. These garments have inbuilt microswitches that are activated by humidity to selectively change porosity across humidities [178].

The attractiveness of spider silk for smart material development lies in its unique mechanical, chemical, thermal, and piezoelectric properties, along with its so-called "shape memory" or how it returns to its original shape upon being deformed. Moreover, many of these properties are variable and adaptable across different environments. Understanding these properties across the entire structural hierarchy, that is from the sub-molecular to macroscopic scale, of the silk fibre is imperative for building materials that vary in mechanical, chemical, thermal, and pleizoelectric property and can be effectively woven into wearable fabrics and function as thermal, stress, or chemical or pollutant sensors.

There are not any spider-silk inspired smart materials currently, or soon to be, on the market but the advancements in recombinant and spinning technologies, along with our increased understanding of the associations between structure and function across the hierarchical levels, means we should see products becoming available at some time in the future.

A property of silk that has made it attractive for fabric making over millennia is its ability to hold natural and synthetic dyes. Indeed, adding various dyes to the mulberry leaves on which silkworms feed induces silkworms to produce cocoons of particularly vibrant colour, depending on the dye ingested. This seems to suggest that if silkworms take up chemicals from the environment they are readily incorporated into their silks.

Among the implications of silk's natural dyeing abilities is that it may be plausible to find ways to induce colouration in fabrics without using highly polluting synthetic dyes and resins. This accordingly represents a way for textile manufacturers to significantly reduce the pollution imprint currently tarnishing the reputation of the fashion and apparel industry. It also has environmental applications, as silkworm cocoons may be used as indicators of chemical pollution[179]. A research team from Japan, for instance, has created transgenic silkworm cocoons that emits green, red or orange fluorescent light in different environments by introducing genes that promote the generation of fluorescent proteins into silkworm eggs. The green fluorescence was achieved by

---

* Examples of existent smart materials include UV sensitive eyeglass lenses, and chemical, light, or moisture sensitive wearable indicators.

inserting genes that had been extracted from jellyfish using transgenic technologies. Red and orange fluorescence on the other hand was achieved using genes that had been extracted from coral[180].

Another likely future application of a manufactured silk-like product in wearables could be the integration of silk's natural properties into health-monitoring clothing. For instance, garments could be designed to incorporate spider silk's piezoelectric properties to monitor and respond to physiological changes in the wearer, such as muscle contractions, heart rate, or even subtle changes in skin temperature. These smart textiles could be used to create clothing that functions as a second skin, capable of detecting early signs of health issues like dehydration, muscle fatigue, or cardiovascular irregularities, and then relaying that information in real-time to a connected device or healthcare provider. Meaning, your health tracker could be a part of your body in the future. Additionally, by harnessing silk's biocompatibility and biodegradability, wearables such as these could be utilized as medical applications, such as bandages that deliver medication while also monitoring the healing process, or implants that degrade safely after serving their purpose. The silk's ability to naturally dye could also allow for the creation of colour-changing fabrics that signal changes in the wearer's health or exposure to harmful environmental conditions, providing both aesthetic and functional benefits in a single innovative garment.

A shift from thermoplastics to silk-like materials in textiles could significantly reduce the amount of microplastics released into the environment. Currently, tiny microplastic particles are being shed from all synthetic fibres all the time and inundate every ecosystem on Earth—from deep oceans to high mountains. These particles have become deeply embedded in food chains, leading to widespread contamination that ultimately affects human health. By replacing thermoplastics with biodegradable silk-like materials, the garment industry could drastically reduce its contribution to this kind of congesting pollution. It will, of course, be the companies that invest early in research and development by experimenting now with natural silks that will gain the edge in developing these products, and thus take a greater slice of the future market in these materials.

Anti-ballistics: silk that defies bullets and disbelief

Of the earmarked applications of spider silk or spider silk-inspired fibres, the one generating the greatest hype, at least among the mainstream media and public at large, is the implied potential to build highly efficient bullet proof vests. As stated already, spider MA silk has superior toughness to Kevlar, the material from which bullet proof vests are currently made, while it comes at a much lower density and so weighs a lot less. This has led the media and many reputable, as well as non-reputable, commentators to conclude that when we reach the stage that we can produce materials mimicking spider silk's properties it will be possible to make tougher, and more light-weight and more durable, impact and ballistic-proof clothing.

I am not sure how far along efforts to create silk-inspired ballistic-proof clothing is, so I cannot comment on whether or not spider silk fibres, or spider silk-inspired fibres, will eventually be used to create bullet proof vests.

I expect that defence departments the world over have been exploring the possibility extremely thoroughly, and that the fact that a technology platform has not yet been developed means it must not be feasible at the moment. However, given the recent technological advances feeding current research into new fibres, I would not rule out the prospect of it eventually becoming possible. The high extensibility of spider MA silk, and its tendency to shrink on exposure to water, may render any direct use of spider silk-like materials problematic in clothing. Currently, it seems like hybridized silks or silks reinforced with carbon fibres or nanotubes might be a more suitable candidate material.

Some media are espousing the future development of an "artificial human skin" that has anti-ballistic properties. This may have been inspired by images on the internet of a visual artist from the Netherlands firing live rounds at a transgenic spider silk block, which was incidentally derived from the Utah State University goat farm, infused with human skin. What was most remarkable about this demonstration was that the silk infused skin did significantly slow down the bullets, albeit bullets fired at a much lower velocity than might be the case in combat. The images were created for an artistic demonstration of the properties of spider MA silk, and not part of any real effort to develop a bullet proof skin, however.

Still, some outlets speculated about the development of bullet-proof soldiers or indestructible android terminators with a spider silk skin over metal endoskeleton (like the Terminator). While these ideas are quite far-fetched now and mostly made for the sake of having some fun, you never know what might be possible in the future. A prospect that is both exciting and scary at the same time.

Winding it up: a silky story for the ages

Here I have covered the most immediately applicable and practical uses of silks, especially spider silks and biomimetic silk-like products, that might be developed in the near future. The range of potential applications for silks is limited only by the extent of one's imagination. Electrospinning of fibroins and spidroins enables the formation of coatings, films, gels, sponges, foams, and powders that can be made into an array of purpose-built functional materials, including ultrathin lenses and smash resistant surfaces, optical imaging devices, fibre-optics, microscope, and camera lens[181]. They may be developed into tough, wear-resistant food and goods packaging materials. They may, if they can be developed in sufficient quantity, offer an environmentally friendly replacement for thermoplastics. Their combination of ultra-strength and light weight could also be harnessed to create gentler and longer lasting car airbags and other such impact resistant material.

Silk-based fibres can be incorporated into wearable yarns and threads. These may be hybridized with materials such as carbon nanotubes and nanoparticles and cellulose-derived particles and fibres to improve the fabric's durability or enhance its conductive or optical properties, or to act as sensors or other micro-electronic devices.

Three-dimensional printing is opening up new possibilities for the practical incorporation of silks into new products, from new fibres, strings, and ropes, to sutures, and prosthetic body parts and supports. There may be a time in the future when enough material can be spun or printed to be used for more broader engineering applications, perhaps as building or bridge scaffolds or supports, or even as a material enabling the building of the ever-speculative space elevator.

Silk-like fibres may be used as robotic moving parts. The ability of spider MA silk to contract and relax under certain stimuli and/or when wet could be harnessed to function as a muscle mimetic in humanoid or animal-inspired robots[182]. Maybe the highly contractile fibres can be built into silk-inspired fabrics or an artificial skin to produce high performance combat robots or android terminators, or switches within smart clothing, or built into spiderman suits.

The potential applications of spider silk extend far beyond traditional uses, delving into realms that seem like science fiction but are increasingly plausible given silk's extraordinary properties.

I imagine a future where spider silk is used to create self-healing clothing that can repair itself after being torn, or camouflage fabrics that can adapt their colour and texture to blend seamlessly with different environments. Spider silk's natural biocompatibility could lead to its integration into advanced medical devices, such as silk-based implants that are both strong and flexible, capable of supporting tissue regeneration or even acting as scaffolds for growing artificial organs. In the aerospace industry, the silk's lightweight and durable nature could inspire the development of ultra-strong, yet feather-light, materials for aircraft or space exploration, perhaps even contributing to the construction of materials for interplanetary missions where saving on weight while not on strength or functionality are of utmost importance.

The piezoelectric properties of spider silk might also lead to the creation of wearable energy-harvesting fabrics that generate electricity from the wearer's movements, powering personal electronics on the go. These futuristic applications, while highly imaginative at the moment, are grounded in the unique capabilities of spider silk and all represent the next exciting frontier in material science innovation.

References

1. Dawkins, R. (1982). *The Extended Phenotype: The Long Reach of the Gene*. Oxford, Oxford University Press.

2. Blamires, S. J., et al. (2017). Physicochemical property variation in spider silk: ecology, evolution, and synthetic production. *Annual Review of Entomology* **62**: 443-460.

3. Morgan, E. (2016). *Gossamer Days: Spiders, Humans, and their Threads*. London, Strange Attractor Press.

4. Darwin, C. (1859). *On the origin of species by means of natural selection, or, The preservation of favoured races in the struggle for life*. London: J. Murray.

5. Craig, C. L. (1997). Evolution of arthropod silks. *Annual Review of Entomology* **42**: 231-267.

6. Malay, A. D., et al. (2017). Analysis of repetitive amino acid motifs reveals the essential features of spider dragline silk proteins. *PLoS One* **12**: e0183397.

7. Finnigan, W. et al. (2020). The effect of terminal globular domains on the response of recombinant mini-spidroins to fiber spinning triggers. Scientific Reports **10**: 10671.

8. McGill, M., et al. (2018). Experimental methods for characterizing the secondary structure and thermal properties of silk proteins. *Macromolecular Rapid Communications* **2018**: 1800390.

9. Jonson, M. A. and D. C. Martin (1999). Finite element modeling of banded structures in *Bombyx mori* silk fibres. *International Journal of Biological Macromolecules* **24**: 139-144.

10. Guinea, G. V., et al. (2006). Volume constancy during stretching of spider silk. *Biomacromolecules* **7**: 2173-2177.

11. Koh, L. D., et al. (2015). Structures, mechanical properties and applications of silk fibroin materials. *Progress in Polymer Science* **46**: 86-110.

12. Little, D. J. and D. M. Kane (2016). Investigating the transverse optical structure of spider silk micro-fibers using quantitative optical microscopy. *Nanophotonics* **6**: 341-348.

13. Blamires, S. J., et al. (2019). Spider silk colour covaries with thermal properties but not protein structure. *Journal of the Royal Society Interface* **16**: 20190199.

14. Yukuhiro, K. et al. (2016) Insect silks and cocoons: structural and molecular aspects. In *Extracellular Composite Matrices in Arthropods*. E. Cohen and B. Moussian eds. Cham, Springer, pp. 515-555.

15. Huang, X., et al. (2012). New secrets of spider silk: exceptionally high thermal conductivity and its abnormal change under stretching. *Advanced Materials* **24**: 1482-1486.

16. Haritos, V. S., et al. (2010). Harnessing disorder: onychophorans use highly unstructured proteins, not silks, for prey capture. *Proceeding of the Royal Society of London B* **277**: 3255–3263.

17. Baer, A. *et al.* (2017). Mechanoresponsive lipid-protein nanoglobules facilitate reversible fibre formation in velvet worm slime. *Nature Communications* **8**: 974.

18. Eliot, S. et al. (1993). A pheromonal function for the crural glands of the onychophoran *Cephalofovea tomahmontis* (Onychophora: Peripatopsidae). *Journal of Zoology* 231: 1-9.

19. Busse, S., et al. (2015). The spinning apparatus of webspinners –functional-morphology, morphometrics and spinning behaviour. *Scientific Reports* **5**: 9986.

20. Edgerly, J.S., et al. (2012). Spinning behaviour and morphology of the spinning glands in male and female Aposthonia ceylonica (Enderlein, 1912) (*Embioptera*: *Oligotomidae*). *Zoologischer Anzeiger* **251**: 297– 306.

21. Collin, M. A., et al. (2009). Comparison of Embiopteran silks reveals tensile and structural similarities across taxa. *Biomacromolecules* **10**: 2268–2274.

22. Piorkowski, D., et al. (2018). Humidity-dependent mechanical and adhesive properties of *Arachnocampa tasmaniensis* capture threads. *Journal of Zoology* **305**: 256-266.

23. von Byern, J., et al. (2017). *Examples of bioadhesives for defence and predation*. Cham, Switz., Springer.

24. Walker, A. A., et al. (2015). More than one way to spin a crystallite: multiple trajectories through liquid crystallinity to solid silk. *Proceedings of the Royal Society B* **282**: 20150259.

25. Wolff, J.O., et al. (2021). Adhesive compounds in the capture threads of glow worms. *Frontiers in Mechanical Engineering* 7: 661422.

26. Wolff, J.O, et al. 2023. Organic salt compositions of pressure sensitive adhesives produced by spiders. Frontiers in Ecology & Evolution 11:1123614.

27. Piorkowski, D. et al. (2021). Structure elongation during stretching in glow worm threads. *Molecules* 26: 3500.

28. Walker, A. A., et al. (2012). Silk from crickets: a new twist on spinning. *PLoS One* 7: e30408.

29. CSIRO Australia (2007). Bees are the new silkworms. *ScienceDaily*, 27 November 2007.

30. Pauling, L., and R. B Corey (1950). Two hydrogen bonded spiral configurations of the polypeptide chain. *Journal of the American Chemical Society* **72**: 534.

31. Crick, F. H. C. (1952). Is α-keratin a coiled coil? *Nature* **170**: 882-883.

32. Kameda, T. (2015). Influence of pH, temperature, and concentration on stabilization of aqueous hornet silk solution and fabrication of salt-free materials. *Biopolymers* **103**: 41-52.

33. Kameda, T., et al. (2012). Identification of hornet silk gene with a characteristic repetitive sequence in *Vespa simillima xanthoptera*. *Comparative Biochemistry and Physiology, Part B* **161**: 17-24.

34. Quicke, D. L. J. and M. R. Shaw (2004). Cocoon silk chemistry in parasitic wasps (Hymenoptera, Ichneumonoidea) and their hosts. *Biological Journal of the Linnean Society* **81**: 161-170.

35. Korenko, S. and S. Pekar (2011). A parasitoid wasp induces overwintering behaviour in its spider host. *PLoS One* **6**: e24628.

36. Addison, J. B., et al. (2013). β-Sheet nanocrystalline domains formed from phosphorylated serine-rich motifs in caddisfly larval silk: a solid state NMR and XRD study. *Biomacromolecules* **14**: 1140-1148.

37. Stewart, R. J. and C. S. Wang (2010). Adaptations of caddisfly larval silk to aquatic habitats by phosphorylation of H-fibroin serines. *Biomacromolecules* **11**: 969-974.

38. Addison, J. B., et al. (2014). Reversible assembly of β-sheet nanocrystals within caddisfly silk. *Biomacromolecules* **15**: 1269–1275.

39. Okano, J., et al. (2016). The effects of surface roughness of sediment particles on the respiration of case-bearing caddisfly larvae. *Freshwater Science* **35**: 611-618.

40. Bauer, F., et al. (2012). Dependence of mechanical properties of lacewing egg stalks on relative humidity. *Biomacromolecules* **13**: 3730-3735.

41. Eisner, T. et al. (1996). Chemical egg defense in a green lacewing (*Ceraeochrysa smithi*). *Proceedings of the National Academy of Science, USA* **93:** 3280-3283.

42. Lintz, E. S. and T. R. Scheibel (2013). Dragline, egg stalk and byssus: a comparison of outstanding protein fibers and their potential for developing new materials. *Advanced Functional Materials* **23**: 4467-4482.

43. Sutherland, T. D., et al. (2014). Convergently-evolved structural anomalies in the coiled coil domains of insect silk proteins. *Journal of Structural Biology* **186**: 402-411.

44. Waite, J. H., et al. (2005). Mussel adhesion: finding the tricks worth mimicking. *Journal of Adhesion* **81**: 297-317.

45. Gantayet, A., et al. (2013). Byssal proteins of the freshwater zebra mussel, *Dreissena polymorpha*. *Biofouling* **29**: 77-85.

46. Asakura, T., et al. (2013). Elucidating silk structure using solid-state NMR. *Soft Matter* **9**: 11440-11450.

47. Kunz, R. I., et al. (2016). Silkworm sericin: properties and biomedical applications. *BioMed Research International* **2016**: 8175701.

48. Yoshioka, T., et al. (2019). A study of the extraordinarily strong and tough silk produced by bagworms. *Nature Communications* **10**: 1469.

49. Kono, N., et al. (2019). The bagworm genome reveals a unique fibroin gene that provides high tensile strength. *Communications Biology* **2**: 148.

50. Craig, H.C, et al. 2022. Nanovoid formation induces property variation within and across individual silkworm silk threads. Journal of Materials Chemistry B. 10:5561-5570.

51. Joel, A. C. and W. Baumgartner (2017). Nanofibre production in spiders without electric charge. *Journal of Experimental Biology* **220**: 2243-2249.

52. Blamires, S. J., et al. (2020). Spider silk biomimetics programs to inform the development of new wearable technologies. *Frontiers in Materials* **7**: 29.

53. Kono, N., et al. (2019). Orb-weaving spider *Araneus ventricosus* genome elucidates the spidroin gene catalogue. *Scientific Reports* **9**: 8380.

54. Babb, P. L., et al. (2017). The *Nephila clavipes* genome highlights the diversity of spider silk genes and their complex expression. *Nature Genetics* **49**: 895-903.

55. Garb, J. E., et al., (2019). The transcriptome of Darwin's Bark Spider glands predicts proteins contributing to dragline silk toughness. *Communications Biology* **2**: 275.

56. Arakawa, K., et al. 2022. 1000 spider silkomics: linking sequence to silk mechanical property. *Science Advances* 41: eabo6043.

57. Craig, H. C., et al. (2019). DNP NMR spectroscopy reveals new structures, residues and interactions in wild spider silks. *Chemical Communications* **55**: 4687-4690.

58. Craig, H. C., et al. (2020). Meta-analysis reveals materiomic relationships in major ampullate silk across the spider phylogeny. *Journal of the Royal Society Interface* **17**: 20200471.

59. Blackledge, T. A., et al. (2012). Biomaterial evolution parallels behavioral innovation in the origin of orb-like spider webs. *Scientific Reports* **2**: 833.

60. Blamires, S. J., et al. (2012). Variation in protein intake induces variation in spider silk expression. *PLoS One* **7**: e31626.

61.   Lacava, M., et al. (2018). Web building and silk properties functionally covary among species of wolf spider. *Journal of Evolutionary Biology* **31**: 968-978.

62.   Blamires, S. J., et al. (2010). Prey type, vibrations and handling interactively influence spider silk expression. *Journal of Experimental Biology* **213**: 3906-3910.

63.   Madsen, B., et al. (1999). Variability in the mechanical properties of spider silks on three levels: interspecific, intraspecific and intraindividual. *International Journal of Biological Macromolecules* **24**: 301-306.

64.   Pérez -Rigueiro, J., et al. (2005). The effect of spinning forces on spider silk properties. *Journal of Experimental Biology* **208**: 2633-2639.

65.   Blamires, S. J., et al. (2013). Uncovering spider silk nanocrystalline variations that facilitate wind-induced mechanical property changes. *Biomacromolecules* **14**: 3484-3490.

66.   Blamires, S. J., et al. (2015). Mechanical performance of spider silk is robust to nutrient-mediated changes in protein composition. *Biomacromolecules* **16**: 1218-1225.

67.   Andersson, M., et al. (2016). Silk Spinning in silkworms and spiders. *International Journal of Molecular Sciences* **17**: 1290.

68.   Parent, L. R., et al. (2018). Hierarchical spidroin micellar nanoparticles as the fundamental precursors of spider silks. *Proceedings of the National Academy of Science, USA* **115**: 11507-11512.

69.   Romer, L. and T. Scheibel (2008). The elaborate structure of spider silk: structure and function of a natural high performance fiber. *Prion* **2**: 154-161.

70.   Andersson, M., et al. (2014). Carbonic anhydrase generates CO2 and H^+ that drive spider silk formation via opposite effects on the terminal domains. *PLoS Biology* **12**: e1001921.

71.   Piorkowski, D., et al. (2018). Ontogenetic shift toward stronger, tougher silk of a web-building, cave-dwelling spider. *Journal of Zoology* **304**: 81-89.

72.   Hagn, F., et al. (2010). A conserved spider silk domain acts as a molecular switch that controls fibre assembly. *Nature* **465**: 239-242.

73. Davies, G. J. G., et al. (2013). Structure and function of the major ampullate spinning duct of the golden orbweaver, *Nephila edulis*. *Tissue and Cell* **45**: 306-311.

74. Keten, S. and M. J. Buehler (2010). Atomistic model of the spider silk nanostructure. *Applied Physics Letters* **96**: 153701.

75. Shao, Z. Z. and F. Vollrath (1999). The effect of solvents on the contraction and mechanical properties of spider silk. *Polymer* **40**: 1799-1806.

76. Elices, M., et al. (2011). Polymeric fibers with tunable properties: lessons from spider silk. *Materials Science and Engineering C* **31**: 1184-1188.

77. Madurga, R., et al. (2015). Persistence and variation in microstructural design during the evolution of spider silk. *Scientific Reports* **5**: 14820.

78. Blamires, S.J. et al. (2023). The spider silk standardization initiative (S3I): A powerful tool to harness biological variability and to systematize the characterization of major ampullate silk fibers spun by spiders from suburban Sydney, Australia. *Journal of the Mechanical Behaviour of Biomedical Materials* 140: 105729.

79. Blamires, S. J., et al. (2012). Environmentally induced post-spin property changes in spider silks: influences of web type, spidroin composition and ecology. *Biological Journal of the Linnean Society* **106**: 580-588.

80. Osaki, S. (2004). Ultraviolet rays mechanically strengthen spider's silks. *Polymer Journal* **36**: 657-660.

81. Colgin, M. A. and R. V. Lewis (1998). Spider minor ampullate silk proteins contain new repetitive sequences and highly conserved non-silk like "spacer regions". *Protein Science* **7**: 667-672.

82. Papadopoulos, P., et al. (2009). Similarities in the structural organization of major and minor ampullate spider silk. *Macromolecular Rapid Communications* **30**: 851–857.

83. Guinea, G. V., et al. (2012). Minor ampullate silks from *Nephila* and *Argiope* spiders: tensile properties and microstructural characterization. *Biomacromolecules* **13**: 2087-2098.

84. Rezac, M., et al. (2017). Morphological and functional diversity of minor ampullate glands in spiders from the superfamily Amaurobioidea (Entelegynae: RTA clade). *Journal of Arachnology* **45**: 198-208.

85. Lee, K. S., et al. (2007). Molecular cloning and expression of the C-terminus of spider flagelliform silk protein from *Araneus ventricosus*. *Journal of Bioscience* **32**: 705-712.

86. Becker, N., et al. (2003). Molecular nanosprings in spider capture-silk threads. *Nature Materials* **2**: 278-283.

87. dos Santos-Pinto, J. R. A., et al. (2018). Spider silk proteome provides insight into the structural characterization of *Nephila clavipes* flagelliform spidroin. *Scientific Reports* **8**: 14674.

88. Vollrath, F., et al. (1990). Compounds in the droplets of the orb spider's viscid spiral. *Nature* **345**: 526-528.

89. Sahni, V., et al. (2014). Prey capture adhesives produced by orb-weaving spiders. In *Biotechnology of Silk*. T. Asakura and T. Miller eds. Dordrecht, Springer, pp. 203-217.

90. Opell, B. D. and M. L. Hendricks (2009). The adhesive delivery system of viscous capture threads spun by orb-weaving spiders. *Journal of Experimental Biology* **212**: 3026-3034.

91. Blamires, S. J. (2019). Biomechanical costs and benefits of sit-and-wait foraging traps. Israel Journal of Ecology and Evolution 66:1-10.

92. Opell, B. D., et al. (2018). Tuning orb spider glycoprotein glue performance to habitat humidity. *Journal of Experimental Biology* **221**: jeb161539.

93. Blamires, S. J., et al. (2014). Nutrient deprivation induces property variations in spider gluey silk. *PLoS One* **9**: e88487.

94. Sahni, V., et al. (2011). Changes in the adhesive properties of spider aggregate glue during the evolution of cobwebs. *Scientific Reports* **1**: 41.

95. Blamires, S. J., et al. (2017). Diet-induced co-variation between architectural and physicochemical plasticity in an extended phenotype. *Journal of Experimental Biology* **220**: 876-884.

96. Blamires, S. J., et al. (2018). Fitness consequences of extended phenotypic plasticity. *Journal of Experimental Biology* **221**: jeb.167288.

97. Townley, M. A., et al. (2006). Changes in composition of spider orb web sticky droplets with starvation and web removal and synthesis of sticky droplet compounds. *Journal of Experimental Biology* **209**: 1463-1486.

98. Opell, B. D., et al. (2017). Humidity-mediated changes in an orb spider's glycoprotein adhesive impact prey retention time. *Journal of Experimental Biology* **220**: 1313-1321.

99. Garb, J. E., et al. (2006). Silk genes support the single origin of orb webs. *Science* **312**: 1762.

100. Collin, M. A., et al. (2016). Evidence from multiple species that spider silk glue component ASG2 is a spidroin. *Scientific Reports* **6**: 21589.

101. Meyer, A., et al. (2014). Compliant threads maximize spider silk connection strength and toughness. *Journal of the Royal Society Interface* **11**: 20140561.

102. Sahni, V., et al. (2012). Cobweb-weaving spiders produce different attachment discs for locomotion and prey capture. *Nature Communications* **3**: 1106.

103. Wolff, J. O., et al. (2015). Spider's super-glue: thread anchors are composite adhesives with synergistic hierarchical organization. *Soft Matter* **11**: 2394-2403.

104. Blasingame, E., et al. (2009). Pyriform spidroin 1, a novel member of the silk gene family that anchors dragline silk fibers in attachment discs of the black widow spider, *Latrodectus hesperus. Journal of Biological Chemistry* **284**: 29097-29108.

105. Geurts, P., et al. (2010). Synthetic spider silk fibers spun from pyriform spidroin 2, a glue silk protein discovered in orb-weaving spider attachment discs. *Biomacromolecules* **11**: 3495-3503.

106. Chaw, R. C., et al. (2017). Complete gene sequence of spider attachment silk protein (PySp1) reveals novel linker regions and extreme repeat homogenization. *Insect Biochemistry and Molecular Biology* **81**: 80-90.

107. Ayoub, N. A., et al. (2013). Ancient properties of spider silks revealed by the complete gene sequence of the prey-wrapping silk protein (AcSp1). *Molecular Biology and Evolution* **30**: 589-601.

108. Addison, B., et al. (2018). Spider prey-wrapping silk is an α-helical coiled-coil/β-sheet hybrid nanofiber. *Chemical Communications* **54**: 10746-10749.

109. Shimanovich, U., et al. (2018). Biophotonics of native silk fibrils. *Macromolecular Bioscience* **2018**: 1700295.

110. Herberstein, M. E., et al. (2000). The functional significance of silk decorations of orb-web spiders: a critical review of the empirical evidence. *Biological Reviews* **75**: 649-669.

111. Cheng, R. C., et al. (2010). Insect form vision as one potential shaping force of spider web decoration design. *Journal of Experimental Biology* **213**: 759-768.

112. Yeh, C. W., et al. (2015). Top down and bottom up selection drives variations in frequency and form of a visual signal. *Scientific Reports* **5**: 9543.

113. Liu, M. H., et al. (2014). Evidence of bird dropping masquerading by a spider to avoid predators. *Scientific Reports* **4**: 5058.

114. Tseng, H. J., et al. (2011). Trap barricading and decorating by a well-armored sit-and-wait predator: extra protection or prey attraction? *Behavioral Ecology and Sociobiology* **65**: 2351-2359.

115. Chaw, R. C., et al. (2016). Candidate egg case silk genes for the spider *Argiope argentata* from differential gene expression analyses. *Insect Molecular Biology* **25**: 757-768.

116. Garb, J. E. and C. Y. Hayashi (2005). Modular evolution of egg case silk genes across orb-weaving spider superfamilies. *Proceedings of the National Academy of Sciences, USA* **102**: 11379-11384.

117. Hu, X. W., et al. (2005). Egg case protein 1. A new class of silk proteins with fibroin-like properties from the spider *Latrodectus hesperus*. *Journal of Biological Chemistry* **280**: 21220–21230.

118. Alam, P., et al. (2016). The toughest recorded spider egg case silks are woven into composites with tear-resistant architectures. *Materials Science and Engineering C* **69**: 195-199.

119. Hu, X. W., et al. (2005). Araneoid egg case silk: a fibroin with novel ensemble repeat units from the black widow spider, *Latrodectus hesperus*. *Biochemistry* **44**: 10020-10027.

120. Barrantes, G. et al. (2013). Variation and possible function of egg sac coloration in spiders. Journal of Arachnology **41**: 342-348.

121. Joel, A. C., et al. (2015). Cribellate thread production in spiders: complex processing of nanofibres into a functional capture thread. *Arthropod Structure and Development* **44**: 568-573.

122. Guo, C., et al. (2018). Structural comparison of various silkworm silks: an insight into the structure–property relationship. *Biomacromolecules* **19**: 906-917.

123. Lai, C. W., et al. (2017). A trap and a lure: dual function of a nocturnal animal construction. *Animal Behaviour* **130**: 159-164.

124. Correa-Garhwal, S. M., et al. (2018). Silk genes and silk gene expression in the spider *Tengella perfuga* (Zoropsidae), including a potential cribellar spidroin (CrSp). *PLoS One* **13**: e0203563.

125. Bai, H., et al. (2012). Functional fibers with unique wettability inspired by spider silks. *Advanced Materials* **24**: 2786-2791.

126. Piorkowski, D. and T. A. Blackledge (2017). Punctuated evolution of viscid silk in spider orb webs supported by mechanical behavior of wet cribellate silk. *Science of Nature* **104**: 67.

127. Fernandez, R., et al. (2014). Phylogenomic analysis of spiders reveals nonmonophyly of orb weavers. *Current Biology* **24**: 1772-1777.

128. Tokareva, O., et al. (2013). Recombinant DNA production of spider silk proteins. *Microbial Biotechnology* **6**: 651-663.

129. Prince, J. T. et al., (1995). Construction, cloning, and expression of synthetic genes encoding spider dragline silk. *Biochemistry* **34**: 10879-10885.

130. Scheibel, T. (2004). Spider silks: recombinant synthesis, assembly, spinning, and engineering of synthetic proteins. *Microbial Cell Factories* **3**: 14.

131. Yang, Y. X., et al. (2016). Hyper-production of large proteins of spider dragline silk MaSp2 by *Escherichia coli* via synthetic biology approach. *Process Biochemistry* **51**: 484–490.

132. Xia, X. X., et al. (2010). Native-sized recombinant spider silk protein produced in metabolically engineered *Escherichia coli* results in a strong fiber. *Proceedings of the National Academy of Science, USA* **107**: 14059-14063.

133. Grip, S., et al. (2009).Engineered disulfides improve mechanical properties of recombinant spider silk. *Protein Science* **18**: 1012-1022.

134. Boulet-Audet, M., et al. (2019). Methods of generating recombinant spider silk protein fibers. Emeryville, CA: U.S. Patent Application 16/176, 939: 1–55

135. Heidebrecht, A. and T. R. Scheibel (2013). Recombinant production of spider silk proteins. *Advances in Applied Microbiology* **82**: 115-152.

136. Pham, T., et al. (2014). Dragline silk: a fiber assembled with low-molecular-weight cysteine-rich proteins. *Biomacromolecules* **15**: 4073-4081.

137. Simmons, J. R., et al. (2019). Recombinant pyriform silk fiber mechanics are modulated by wet-spinning conditions. *ACS Biomaterial Science and Engineering* **5**: 4985–4993.

138. Barr, L. A., et al. (2004). Production and purification of recombinant DP1B silk -like protein in plants. *Molecular Breeding* **13**: 345–356.

139. Scheller, J., et al. (2001). Production of spider silk proteins in tobacco and potato. *Nature Biotechnology* **19**: 573–577.

140. Rising, A., et al. (2011). Spider silk proteins: recent advances in recombinant production, structure–function relationships and biomedical applications. *Cellular and Molecular Life Sciences* **68**: 169-184.

141. Zhang, X., et al. (2019). CRISPR/Cas9 initiated transgenic silkworms as a natural spinner of spider silk. *Biomacromolecules* **20**: 2252–2264

142. Butler, D. (2014). The silk route. *The Biologist* **61**: 12-15.

143. Copeland, C. G., et al. (2015). Development of a process for the spinning of synthetic spider silk. *ACS Biomaterial Science and Engineering* **1**: 577-584.

144. International Silkworm Genome Consortium (2008). The genome of a lepidopteran model insect, the silkworm *Bombyx mori*. *Insect Biochemistry and Molecular Biology* **38**: 1036-1045.

145. Garb, J. E., et al. (2019). The transcriptome of Darwin's bark spider silk glands predicts proteins contributing to dragline silk toughness. *Communications Biology* **2**: 275.

146. Venkatesan, H., et al. (2019). Artificial spider silk is smart like natural one: having humidity-sensitive shape memory with superior recovery stress. *Materials Chemistry Frontiers* **3**: 2472.

147. Andersson, M., et al. (2017). Biomimetic spinning of artificial spider silk from a chimeric minispidroin. *Nature Chemical Biology* **13**: 262-264.

148. Gu, L., et al. (2019). Scalable spider-silk-like supertough fibers using a pseudoprotein polymer. *Advanced Materials* **31**: 1904311.

149. Corsini, P., et al. (2007). Influence of the draw ratio on the tensile and fracture behavior of NMMO regenerated silk fibers. *Journal of Polymer Science Part B* **45**: 2568-2579.

150. Yao, Y., et al. (2020). Improving the tensile properties of wet spun silk fibers using rapid Bayesian algorithm. *ACS Biomaterial Science and Engineering* **6**: 3197–3207.

151. Doblhofer, E., et al. (2015). To spin or not to spin: spider silk fibers and more. *Applied Microbiology and Biotechnology* **99**: 9361-9380.

152. Pérez-Rigueiro, J., et al. (2018). Straining flow spinning of artificial silk fibers: a review. *Biomimetics* **3**: 3040029.

153. Magaz, A., et al. (2018). Porous, aligned, and biomimetic fibers of regenerated silk fibroin produced by solution blow spinning. *Biomacromolecules* **19**: 4542–4553.

154. Yoshioka, T., et al. (2016). Molecular orientation enhancement of silk by the hot-stretching- induced transition from αHelix-HFIP complex to βSheet. *Biomacromolecules* **17**: 1437–1448.

155. Albertson, A. E., et al. (2014). Effects of different post-spin stretching conditions on the mechanical properties of synthetic spider silk fibers. *Journal of the Mechanical Behavior of Biomedical Materials* **29**: 225-234.

156. Su, I., et al. (2020). Perspectives on three-dimensional printing of self assembling materials and structures. *Current Opinions in Biomedical Engineering* **15**: 59-67.

157. Zhang, M., et al. (2019). Printable smart pattern for multifunctional energy-management e-textile. *Matter* **1**: 1-12.

158. Debabov, V. G. and V. G. Bogush (2020). Recombinant spidroins as the basis for new materials. *ACS Biomaterial Science and Engineering* **6**: 3745–3761.

159. Holland, C., et al. (2018). The biomedical use of silk: past, present, future. *Advanced Healthcare Materials* **2018**: 1800465.

160. Qi, Y., et al. (2017). A review of structure construction of silk fibroin biomaterials from single structures to multi-level structures. *International Journal of Molecular Sciences* **18**: 237.

161. Guan, G., et al. (2010). Promoted dermis healing from full-thickness skin defect by porous silk fibroin scaffolds (PSFSs) Biomedical Materials and Engineering **20**: 295–308.

162. Kundu, B., et al. (2014). Isolation and processing of silk proteins for biomedical applications. *International Journal of Biological Macromolecules* **70**: 70-77.

163. Schacht, K. and T. R. Scheibel (2014). Processing of recombinant spider silk proteins into tailor-made materials for biomaterials applications. *Current Opinion in Biotechnology* **29**: 62-69.

164. Ebrahimi, D., et al. (2015). Silk–its mysteries, how it is made, and how it is used. *ACS Biomaterials Science and Engineering* **1**: 864-876.

165. Koh, K. S. (2018). Bone regeneration using silk hydroxyapatite hybrid composite in a rat alveolar defect model. *International Journal of Medical Sciences* **15**: 59-68.

166. Yao, D., et al. (2016). Silk scaffolds for musculoskeletal tissue engineering. *Experimental Biology and Medicine* **241**: 238–245.

167. Fan, H., et al. (2009). Anterior cruciate ligament regeneration using mesenchymal stem cells and silk scaffold in a large animal model. *Biomaterials* **30**: 4967-4977.

168. Allardyce, B. J. et al. (2016). Comparative acoustic performance and mechanical properties of silk membranes for the repair of chronic tympanic membrane perforations. *Journal of the Mechanical Behavior of Biomedical Materials* **64**: 65-74.

169. Cao, Y. and B. Wang (2009). Biodegradation of silk biomaterials. *International Journal of Molecular Sciences* **10**: 1514-1524.

170. Brown, J. E., et al. (2016). Thermal and structural properties of silk biomaterials plasticized by glycerol. *Biomacromolecules* **17**: 3911-3921.

171. Lammel, A., et al. (2011). Recombinant spider silk particles as drug delivery vehicles. *Biomaterials* **32**: 2233-2240.

172. Bar-Cohen, Y. (2006). Biomimetics—using nature to inspire human innovation. *Bioinspiration and Biomimetics* **1**: P1-P12.

173. Dams-Kozlowska, H., et al. (2013). Purification and cytotoxicity of tag-free bioengineered spider silk proteins. *Journal of Biomedical Materials Research Part A* **101A**: 456-464.

174. Morikawa, J., et al. (2016). Silk fibroin as a water-soluble bio-resist and its thermal properties. *ACS Advances* **6**: 11863-11869.

175. Sahni, V., et al. (2014). Direct solvation of glycoproteins by salts in spider silk glues enhances adhesion and helps to explain the evolution of modern spider orb webs. *Biomacromolecules* **15**: 1225-1232.

176. Diaz, C., et al. (2020). The moth specialist spider *Cyrtarachne akirai* uses prey scales to increase adhesion. *Journal of the Royal Society Interface* **17**: 20190792.

177. Joel, A. C., et al. (2020). Biomimetic combs as antiadhesive tools to manipulate nanofibers. *ACS Applied Nanomaterials* **3**: 3395-3401.

178. Flanagan, P. et al., (2024). Hybrid sensor configurations. In *Organic and Inorganic Material Based Sensors*. S. Tomas, S. Das, P. Pratim Das, eds. Weinheim, Wiley VCH.

179. Nisal, A., et al. (2014). Uptake of azo dyes into silk glands for production of colored silk cocoons using a green feeding approach. *ACS Sustainable Chemistry and Engineering* **2**: 312-317.

180. Babu, K. M. (2012). Silk production and the future of natural silk manufacture. In *Handbook of Natural Fibres Volume 2: Processing and Applications*. R. M. Kozlowski ed. Oxford, The Textile Institute.

181. Reed, E. J., et al. (2009). Biomimicry as a route to new materials: what kinds of lessons are useful? *Philosophical Transactions of the Royal Society of London A* **367**: 1571-1585.

182. Agnarsson, I., et al. (2009). Spider silk as a novel high performance biomimetic muscle driven by humidity. *Journal of Experimental Biology* **212**: 1990-1994.

www.ingramcontent.com/pod-product-compliance
Lightning Source LLC
Chambersburg PA
CBHW061139030426
42335CB00002B/47